*WrittenInDetroit.com*

Twitter: @WriteDetroit
Facebook: WrittenInDetroit

# THE FIRST MOBILE APP

## BY R.J. KING

A portion of each book sold will be donated to Beyond Basics, a nonprofit organization that teaches students in Detroit Public Schools to read at their grade level or higher within six weeks (95 percent success rate), and to the STEM educational program at the Davis Aerospace Technical High School curriculum in Detroit.

King Publishing Co., Momentum Books

*8TrackBook.com*
*writtenindetroit.com*
*MomentumBooks.com*

THIS BOOK IS DEDICATED TO:

**My Parents**
John and Barbara

**My Brothers and Sisters (and their Spouses)**
Kathy, Linda (Scott), Nancy (Craig), Patrick (Linda), Mary (Brad),
Suzy, Maureen (Marc), and John (Li-Hsing)

**My Nieces and Nephews**
Mary Clare, Maureen Megan, Rosemary, Zachary, James, Kathryn,
Christopher, Nicholas, Stephanie, Crystal, Michael, Melissa,
Matthew, Megan, Amanda, Joshua, Julien, Sophie, Kate, Ian, Matthew, Alice,
and Leon

**Thanks to the following families/firms/organizations:**

| Lear | The Henry Ford | Museum of Flight |
|------|----------------|------------------|
| Kusisto | Martines | Huth Lynett |
| Campbell | Carmona | Brooks Kushman |
| Muntz | Ford Motor Co. | Hour Media |

**And Special Thanks to the Team:**
Josef Bastian (Advisor)
Cassidy Zobl (Designer)
Anne Berry Daugherty (Copy Editor)
Carl Winans (Web Designer)
Stephanie King (Illustrator)

# INTRODUCTION

I grew up in an automotive family in Detroit, and my parents would, at times, host work parties at our home. One of my fondest memories is serving drinks with my brother, Patrick, behind a bar in our basement. Around us, popular music flowed from an 8 Track tape player. There was one rule at these parties: My brother and I could pour water, soda, or juices, but someone older had to handle the big bottles.

What we didn't appreciate at the time was that our Dad, John P. King, who was hired by Ford Motor Co. on January 11, 1965, as the project engineer of the 8 Track tape player, had overseen the development of the first mobile app. The unit, introduced in October 1965, consisted of an AM radio with an integrated 8 Track tape player. At the start, consumers had 175 albums to choose from, but in quick order, every major artist became available (some begrudgingly). Invented by Bill Lear, the rollout was led by four companies: Ford, Motorola, RCA Victor, and Lear Jet Stereo.

After reading through a 1975 technical paper my Dad presented to an industry trade group, the International Tape Association, the full story of the debut of the world's first mobile app came forward. From the outset, the 8 Track was a raw medium, but before long, it proved to be a popular feature in cars, homes, offices, airplanes, boats, public buildings; even airport terminals.

Going against an existing 4 Track player introduced by Earl Muntz in 1960, Ford sold some 125,000 8 Track tape units over the first model year (1966). Soon thereafter, Chrysler entered the marketplace, followed by General Motors, American Motors, and Volkswagen. In turn, home and office units began appearing in consumer electronics stores. By 1970, it was a billion-dollar industry.

Without the technical advances of magnetic tape and playback units in the early 1960s, along with the breakout of major recording artists and the first wave of baby boomers reaching young adulthood, the 8 Track tape player would never have made it out of the gate.

While today music is available in nearly an instant, there was a time when consumers had just two options: an AM radio or a record player. The 8 Track player was, in modern parlance, a disruptor, despite the fact that several record companies flat-out dismissed it as a passing fad. This is a story of how a small group of people broke away from the status quo and ushered in a remarkable playback system that expanded the availability of music to nearly every corner of the world.

*—R.J. King*

# CONTENTS

# INSIDE TRACK

**BILL LEAR**: A prolific inventor with 110 patents to his name, Lear developed the 8 Track tape player in 1964 following a heated exchange with Earl Muntz over the viability of his 4 Track tape player. To get the 8 Track rolling, Lear called Henry Ford II, who agreed that Ford Motor Co. would be the first to offer the player in late 1965 for the 1966 model year. For music, Lear phoned his good friend, RCA Chairman David Sarnoff, who extended 175 albums for the new medium. Launched in October 1965, Ford offered combined AM radio and 8 Track stereo tape players supplied by Motorola, while Lear provided the cartridges. Three months after the launch, in January 1966, Lear Jet Stereo began to supply Chrysler and went on to build and sell 8 Track units and various electronic devices for homes and offices. Lear stayed on as chairman following the sale of his aircraft and electronics operations to Gates Rubber Co. in 1966.

**EARL "MADMAN" MUNTZ**: A marketing whiz and self-taught engineer, Muntz introduced low-priced portable black-and-white televisions in the 1950s, which sold by the thousands at $99 per unit. A successful used-car dealer in Los Angeles, he was the first to develop an alter ego to move the metal. The personas he created in radio and television commercials were widely imitated. Following the production of 320 Muntz Jet cars (complete with a liquor cabinet in the back), he tried his hand at developing portable color TVs. Unable to find the right balance between price and presentation, Muntz debuted an aftermarket 4 Track player in 1960. When 8 Track/AM radio debuted in late 1965, he was the undisputed leader in pre-recorded tape sales, but the 4 Track system faded out in the early 1970s due to intense competition and a lack of upgrades.

**HENRY FORD II**: After rebuilding Ford Motor Co. following World War II — domestic sales were slow to recover, while much of the company's European operations were in shambles — Ford II and the "Whiz Kids" he assembled to supplement his burgeoning CEO skills pushed the company into new horizons. The 1949 Ford all but saved the company, and Ford II initiated a price war in 1953 that led to industry consolidation and market share gains. Still, in 1956 he nixed a plan to install 4 Track tape players developed by Motorola, ostensibly to devote more resources to the launch of the Edsel. When the car line bombed, Ford II was much more selective. Still, he was willing to take a chance on Lear and the 8 Track tape player, which proved to be a major success.

---

**DONALD N. FREY**: A talented engineer who also gained expertise in design, new model introductions, sales, and manufacturing, Frey joined Ford in 1950 and was named vice president and chief engineer in 1964. He oversaw the introduction of the 1964 1/2 Mustang (Lee Iacocca was his boss), as well as the combined AM radio and 8 Track stereo tape player. It was Frey who insisted, along with some of Ford's powerful product planners, that Motorola develop and supply the combined AM radio and tape player — much to Lear's dismay. Early in product development, Frey nixed a plan by Motorola to supply a combined 4 Track and 8 Track player. The move simplified the rollout of the 8 Track for the 1966 model year and set up a fierce competition with Muntz.

---

**DAVID SARNOFF**: A pioneer in broadcasting in the 1920s, Sarnoff was instrumental in establishing radio as a mass medium and oversaw RCA's acquisition of Victor Talking Machine Co., then the largest producer of records. In 1930, following an antitrust suit in which Westinghouse and General Electric had to divest their investment in RCA Victor, Sarnoff became the de facto leader of the company until his retirement in 1970. A good friend of Lear, he agreed to supply the music for the 8 Track — 175 titles at first, before opening up most of the company's musical library. After

presenting Elvis Presley with a gold Cadillac in 1965, Sarnoff arranged to have two 8 Track tape players installed in the vehicle's dashboard by the engineering team of Lear Jet Stereo.

---

**JOHN P. KING**: An expert in magnetic tape and electrical engineering, King was hired by Ford Motor Co. on January 11, 1965, as project engineer of the 8 Track tape player. Racing to deliver the unit in nine months' time, King wrote the specs, oversaw early reliability tests, and shepherded the introduction via frequent trips between Motorola's headquarters in Chicago (Franklin Park) and its radio production facility in Quincy, Ill., as well as RCA's record factory in Indianapolis and Ford's various manufacturing facilities. King, on his own volition, installed upgraded speakers in his 1963 Ford Fairlane station wagon in preparation for a "Sound Off" between the 8 Track and Muntz' 4 Track. King won the contest, thereby cementing the 8 Track as a viable and successful product that began to outpace Muntz' 4 Track system in 1967.

---

**OSCAR P. KUSISTO**: A risk-taker and gifted engineer, Kusisto was a corporate officer at Motorola as well as vice president and general manager of the automotive products division. He led a small team of engineers that delivered a combined AM radio and 8 Track tape player to Ford. Eager to remain a key electronic supplier to the automaker — Ford acquired radio supplier Philco in 1961 — Kusisto stayed one step ahead of the competition by offering push buttons, FM radio, and quadrasonic sound. In the early days of 8 Track, he assumed the development of home units after Motorola's consumer products division balked at the project.

---

**LARRY FINLEY**: A brilliant promoter, Finley established International Tape Cartridge Corp. in New Jersey in 1965. A former nightclub owner in New York, Finley moved to Los Angeles in the 1930s and set up a line of jewelry stores. He went on to produce and host TV and

radio shows before buying up the tape cartridge rights for close to 60 labels, including Dot, Audio-Fidelity, Horizon, Seeco, Tico, and Vee Jay. Finley was instrumental in driving the cartridge business in the United States and Europe via numerous merchant and consumer promotions, as well as a weekly column that ran in *Billboard* throughout the second half of the 1960s. He also was an organizer and promoter of numerous trade and industry conferences, and co-founded the International Tape Association.

**ED CAMPBELL**: As president of Lear Jet Stereo, Campbell joined the Detroit-based electronics supplier in 1966 following the acquisition by Gates Rubber Co. (along with Lear's aircraft operations). He led an effort to redesign the early Lear player, which was too heavy and costly. Overcoming numerous quality problems, the new unit introduced in 1967 proved to be a hit with various automakers, including Chrysler, AMC, and Volkswagen. Campbell also negotiated contracts with suppliers of electric motors based in Japan. While the Gates legal team didn't think much of Lear's patents for the 8 Track tape player, Campbell won several legal showdowns that resulted in millions of dollars in royalties. He went on to establish the International Tape Association with Finley and Kusisto.

**IRWIN TARR**: The slick division vice president of recording tape marketing at RCA Victor in New York was initially content to control the 8 Track, but he soon prodded and cajoled his competitors to support the medium. At first, RCA committed 175 titles to the project; Tarr was reluctant to release more albums until a phone call from his boss, David Sarnoff, changed his mind. As much as RCA Victor would have liked to control the cartridge tape industry, Sarnoff and Tarr realized consumers would lose interest if all of their favorite recording artists weren't available. Tarr worked with automakers, suppliers, distributors, and record stores to promote and advance the 8 Track, and he represented RCA and the industry at conferences and trade shows.

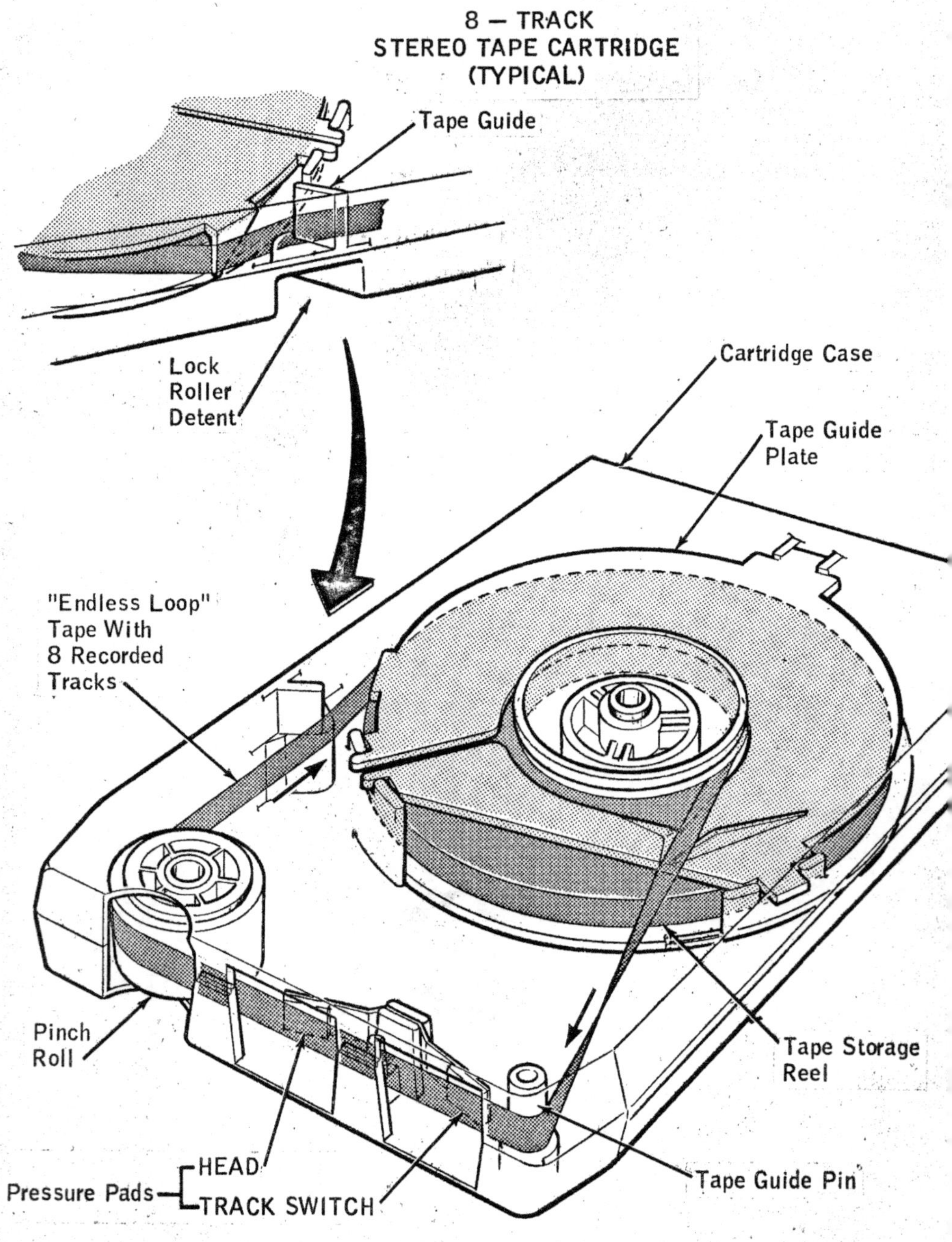

# 8 — TRACK
## STEREO TAPE CARTRIDGE
## (TYPICAL)

Tape Guide

Lock
Roller
Detent

Cartridge Case

Tape Guide
Plate

"Endless Loop"
Tape With
8 Recorded
Tracks

Pinch
Roll

Tape Storage
Reel

Pressure Pads — HEAD
TRACK SWITCH

Tape Guide Pin

DO NOT OPEN CARTRIDGE
This Illustration Is To Be Used For Reference Only

Figure 9

# 01

# BIRTH
# OF
# AN
# INDUSTRY

## MOTOR VICTROLA

WHEN BILL LEAR WAS SEARCHING FOR an entertainment system for his corporate jet program in the early 1960s, his daughter, Shanda, showed him a hybrid version of a tape cartridge system that had been in use by radio stations since 1957. The stereo cartridges — or "carts," as disc jockeys dubbed them — replaced live performances and reel-to-reel recordings of music, jingles, and commercials. The closed-loop tape cartridges, similar in size to a small paperback, offered around five minutes of recording time. Once a cartridge was inserted into a tape player, a built-in pinch roller flipped up and created the necessary pressure to start the audio playback process.

The so-called Fidelipac tapes, developed in 1954 by inventor George Eash (others credit the cartridge system to Bernard Cousino), were a major advancement over the reel-to-reel tape players prevalent in the recording and broadcasting industries. The bulky reel-to-reel players were difficult to use — thin, black tape one-quarter inch in width had to be guided by hand through a series of levers and rollers. They also were expensive, and weighed as much as 20 pounds. As a result, there was limited demand for reel-to-reel players inside American homes and offices.

On the other hand, record players, or phonographs, had enjoyed singular popularity since the 1920s, and again in the 1930s when RCA Victor introduced electronic turntables. Initially offered in large decorative cabinets, over time record players were downsized and made less expensive. In a stable environment, a needle, or stylus, traced the grooves of a record as it spun slowly on a turntable. If it was jarred or bumped, the needle could pop out, or skip, from groove to groove. To an ambitious entrepreneur, the development of a pre-recorded tape system featuring popular music was the holy grail of a new industry — opening a vast consumer market to singers, musicians, engineers, and technicians.

At the dawn of the 1960s, America was ready for a novel, mobile entertainment offering. The first baby boomers — the generation born between 1946 and 1964 —were just coming of age. The Korean War was over, and in 1961 John F. Kennedy had become the youngest president elected to the office. The nation, for the first time since the boom years of the 1920s, was seemingly in control of its destiny. Tapping this new spirit of adventure, millions of Americans began to free themselves of the rigidity of urban life. Starting in the 1950s, families large and small purchased new homes outside of big cities. They wanted a bigger slice of the planet free of factories, pollution, and traffic. Along with a new car or two, an attached garage, and perhaps a swimming pool in the backyard, the wave of suburbanites wanted their homes outfitted with the latest technology. But apart from programming offered by television producers and radio stations, the phonograph was the only way consumers could select the music they wanted, no matter the time. The record player, in essence, afforded consumers a measure of control over the variances and whims of radio stations and disc jockeys. And they sold like hotcakes.

Ever since Columbia Records introduced LP (Long Play) records in 1948, consumers had been able to visit their favorite store to purchase new albums by Frank Sinatra, Bing Crosby, Doris Day, or Rosemary Clooney. Thomas Edison invented the phonograph in 1877 at his Menlo Park Research Laboratory in New Jersey (since moved to Greenfield Village at The Henry Ford in Dearborn, Mich.), jump-starting the nation's love affair with pre-recorded music. Edison used a tin foil cylinder to record sound, but his invention was quickly supplemented by Alexander Graham Bell, whose Graphophone used higher-quality wax cylinders. In 1887, Columbia Phonograph Co. was founded to market various Graphophone products, which, for all intents and purposes, was the official launch of the record industry (the company eventually became Columbia Records). In 1921, for the first time, record sales in the United States topped $100 million in annual revenue, according to the Record Industry Association of America. By 1968, sales surpassed the $1 billion mark.

The introduction of a mobile pre-recorded entertainment player was a foregone conclusion. The only question was whether the record companies would support a new entertainment offering given the associated costs of design, copyrights, materials, production, distribution, and marketing.

The first attempt came in the form of Earl "Madman" Muntz, a used-car dealer in the Los Angeles area who used his promotional skills, technical savvy, and friendly sway within the music industry to develop a 4 Track stereo tape player for use inside automobiles, homes, offices, boats, and planes. He took the original Fidelipac tapes used in radio stations and designed a player that could hold a slightly larger cartridge. The system, which extended the playback time to 16 minutes, owed a debt of gratitude to Motorola, which had developed a 4 Track player in 1956. But due to an economic slowdown, among other factors, Motorola couldn't convince its main customer, Ford Motor Co., to install the players in its vehicles. In addition, the record companies declined to support it.

The Muntz unit, first available in 1960, was mounted beneath a dashboard or placed next to a record player inside a home or office. With the connection of a few wires, the player could be interfaced with most available speakers. At the start, a stretch of 16 minutes was enough room for four or five songs. Had the used-car salesman been able to develop a "Motor Victrola" with 45 minutes of pre-recorded tape — enough to hold an entire album — the recording industry would have licensed labels to Muntz out of the gate.

Not one to be denied, Muntz turned to the popular nightclub and bar scene in and around Los Angeles. In short order, he licensed music from popular bands and acts at a time when rock 'n' roll was coming of age. From the outset, Muntz sold scores of 4 Track tape players and cartridges through his used-car outlets, his Muntz Auto Stereo stores, and various distributors like Lear. It wasn't long before he convinced labels like Dot Records to license music to the 4 Track system, says Muntz' son, Jim. "George Eash and my father worked on putting 4 Track together, and I was convinced it was ready to go," Jim says. "Viking, in Minnesota, built the first 25 to 50 players, but the problem was we didn't have a motor. We contacted Ford Motor Co. and found a motor they were using for convertibles. The challenge was to make (the motor) run at a consistent speed."

## THE DROPOUT

Born in 1902, William (Bill) Powell Lear was raised on a farm in Hannibal, Mo.. When he was 5 years old, he and his mother moved in with relatives, leaving a father who struggled to provide a steady income. It wasn't long before the pair moved to the Chicago area, and Lear attended Kershaw Grammar School. A self-taught engineer, he spent hours at the library reading and studying wireless technology; mechanical systems; and the work of Nicola Tesla, the father of the alternating current (AC) electrical system, and Guglielmo Marconi, who spearheaded long-distance radio transmission, among others. Independent and self-assured, Lear was twice kicked out of high school for showing up his teachers. In one incident, he corrected a mistake in electric class — the teacher had said an ammeter (an instrument used to measure electric current in amperes) only worked when connected to the positive side of a circuit. Lear raised his hand and informed the class that the ammeter measured current regardless of a positive or negative connection. After Lear took it upon himself to demonstrate the connection in front of the class, the teacher promptly kicked him out. He ran into similar resistance in his physics and shop classes, and soon left school. A few years later he came close to finishing high school but, again, he was asked to leave, having showed up a teacher a second time. In 1944, Lear and Moya Olsen, the daughter of Vaudeville comedian John "Ole" Olsen, brought into the world their first daughter, Shanda, who went on to become a noted entertainer, public speaker, and business owner.

Along with various partners, Bill Lear introduced numerous innovations in radio communications, most notably a version of the car radio in the early 1930s. Paul Galvin, of the Galvin Manufacturing Co., and Lear renamed the operation Motorola during a road trip from Chicago to Atlantic City to introduce their car radio at a trade convention. The name was derived from "motor" for the automobile, and "ola" after popular music systems of the day such as Victrola. Lacking the money to set up an exhibit at the Radio Manufacturers Association Convention in Atlantic City, the pair parked Galvin's Studebaker by a pier, added a couple of loudspeakers, and drew enough orders from convention-goers to make the trip a commercial success. "Dad was always scribbling ideas and designs on restaurant napkins and tablecloths, all the while telling jokes and discussing the infinite possibilities of the mind," Shanda says.

The car radio and other innovations generated a substantial income, and Bill used part of the proceeds to purchase a Fleet biplane in 1930 for $2,500 from a woman in Dearborn, Mich. After taking flying lessons, Lear began tinkering with instruments and communications systems for use in the fledgling aviation industry. At the time, pilots would literally fly by the seat of their pants, ever watchful for railroad lines that would help them navigate their way from one town to the next. Over time, Lear almost single-handedly propelled the aviation industry into the modern age by developing radio direction finders, as well as the autopilot and automatic aircraft landing systems.

In the early 1960s, as Lear was developing his business jet in Wichita, Kan., he and his team were looking for a novel entertainment system in which musical selections could be controlled by the user rather than depending on the play lists of disc jockeys. What's more, the resulting Learjet flew so fast and so high that it was impossible to pick up commercial radio signals.

In truth, the 8 Track tape player, and, for that matter, the Learjet and numerous other innovations, would never have come to pass had it not been for Lear's acrophobia, or fear of heights. "It was 1934, and Dad was down on his luck and flat broke," Shanda says. "He had already been married and divorced twice, and he owed alimony and child support to two wives, so he decided to commit suicide." He checked in at the Hotel St. George in New York's Brooklyn Heights neighborhood and rented a suite on the top floor, where he planned to drink a bottle of scotch and then jump. But because of his fear of heights, he couldn't get near the window. "So he drank the bottle, passed out, and woke up the next day with a major hangover," Shanda says.

Connecting with a friend that afternoon, Lear hit upon developing a highly reliable all-wave radio receiver equipped with a turret tuner. His new features would allow for broadcast, low frequency, and three shortwave bands, whether he built it himself or sold the idea to an electronics company. In 1934, all-wave radios were selling well, given users could tune in stations located as many as 5,000 miles away. The trouble was, the early sets required extensive debugging, and no two were exactly alike.

After presenting his idea to an RCA executive, Lear received a check for $500 and went to work. Two weeks later, he showed off his all-wave radio receiver to RCA Victor Division President E.T. Cunningham. "On the way up the elevator at the Empire State Building to meet with Mr. Cunningham, Dad didn't know what his invention was worth," Shanda says. "Cunningham was smart. He kept my dad waiting for an hour, perhaps thinking the delay would unsettle him. During that time, Dad decided to put the onus on Cunningham and make him name a price. Dad had thought of asking for $25,000 but he was willing to take $10,000, and if the offer came for $5,000, he would take that, too. His bargain basement price was $2,500, which would still pay off a lot of his debt."

After being escorted to Cunningham's office, Lear listened as Cunningham relayed the news that RCA had patents on everything the inventor had presented. Lear agreed with him, but noted no one had designed a receiver in a standardized format that would deliver superior quality at a lower cost. Quick to give his potential client the floor, Lear watched as Cunningham leaned back in his chair and said, "OK, Bill, we'll give you $50,000." Shanda says her father was speechless. "Cunningham thought he was trying to drive a harder bargain," she says, "so he offered Dad a $25,000 annual retainer fee for consulting for a period of five years. Still, Dad was stunned. Cunningham threw in another $15,000 worth of business for his company annually for five years. The entire deal was worth $250,000."

In the summer of 1964, after introducing numerous other innovations, Lear — following a disagreement with Muntz over paying a five-cent royalty fee on each 4 Track cartridge — turned his attention to the development of the 8 Track tape player. The inventor noted several drawbacks to the Muntz player, notably limited playing time, a lack of major artists, distribution challenges, and operational deficiencies (users had to manually engage a lever to switch between the two program tracks). In addition, the Muntz 4 Track player included the same internal location for the pinch roller as the Fidelipac system used at radio stations. Lear's design, which he developed with two associates, Frank Schmidt and Sam Auld, placed the pinch roller inside the cartridge, which automated the playback process and doubled the amount of music that could be pre-recorded, among other advances. With the added room for music, full record albums could be pre-recorded on magnetic tape, packaged in a cartridge, and sold to consumers. Prior to the introduction of the LP, a song by an artist or orchestra was recorded and sold as an individual record. The record companies eventually packaged and sold a collection of the individual records as an "album." The name stuck, and to this day a record or compact disc is still referred to as an album.

Because the resulting Lear design was a major upgrade from the 4 Track, the inventor was convinced the automotive industry, namely Ford, would entertain the introduction of a seamless, pre-recorded system featuring a player integrated into the dashboard, or attached just below it. But as Lear unveiled his advancement in late 1964, some executives at Ford were concerned that the 8 Track tape player was problematic and couldn't be readied for a planned launch on October 3, 1965 (for the 1966 model year). Meanwhile, Muntz, who had received word that Lear had accomplished the impossible, took another run at Ford.

"Muntz likely didn't know where we were at with the 8 Track introduction, and we weren't about to tell him," says John P. King, Ford's project engineer for the 8 Track tape player, who joined the automaker on January 11, 1965. "He actually came in through the back door. Muntz was eager to see his 4 Track system become Ford's official aftermarket player that could be installed by dealers. Normally, when you're looking to introduce a product like his, you would go through our radio department. But he went through our parts and accessories group (based at Ford's Wixom Assembly Plant outside of Detroit)."

The plant, one of the largest automotive factories in the world when it opened on April 1, 1957, produced more than 6.6 million vehicles, mostly Lincolns and Thunderbirds, before production was ceased in 2007 due to falling demand. Vehicles that went through final assembly at the 4.7-million-square-foot complex included the Lincoln Continental (1958-2002), Ford Thunderbird (1958-2005), and Lincoln Town Car (1981-2007). It was an ideal setting for what can best be described as a "Sound Off," where the accessories division compared Muntz' 4 Track player to Lear's patented 8 Track tape player (supplied by Motorola). Knowing the stakes were high, King made sure the 8 Track had its best showing. To achieve the ultimate performance, he equipped his blue 1963 Ford Fairlane station wagon with an early-production 8 Track tape player. In addition, he replaced the car's factory speakers with higher-end, six-by-nine-inch speakers that produced better bass response — two speakers in the front door panels and two units in the rear cargo area.

For his part, Muntz showed off the latest offering of his 4 Track system, which he coined Stereo-Pak (in 1962, Muntz sold Muntz Auto Stereo following a disagreement with investors, and soon started Muntz Stereo-Pak). A flamboyant entertainer and entrepreneur, Muntz, born in 1914, got his start in the business world at age 14 when he repaired radios for his father, a radio distributor in Elgin, Ill. At 20 years old, he branched out into used cars; his mother signed the various sales documents until he was 21. He eventually moved to Los Angeles and opened several used-car lots. It was Muntz who came up with the idea of creating an alter ego to sell used cars on television. His performances were so zany and outrageous that stars like Bob Hope and Jack Benny would mimic him. Because cars would fetch higher prices in California, Muntz ran a profitable business. When he got his start in California in 1940, used cars were selling for a premium as the advent of World War II severely crimped new car production. Muntz' strategy was to purchase much cheaper used cars in the Midwest and pay drivers $50 to get the cars to California. He typically doubled his money.

Muntz went on to design and produce the first black-and-white portable TV, which retailed for less than $100 ($99.95). The unit sold well at first, but with the introduction of color television in the mid-1950s, the company eventually filed for bankruptcy. He also developed the first wide-screen projection TV and sold cellular phones, satellite dishes, and large-screen televisions until 1986, when he was diagnosed with lung cancer. In his later years, Muntz drove a white Lincoln Continental that was outfitted with a television in the dashboard. He said the TV helped him "drive better." He passed away in 1987.

Another Muntz project was the development of a sports car he dubbed the Road Jet. Originally crafted by Frank Kurtis, a racecar designer, the two-seater Kurtis Kraft Sport was first offered in 1950 but, given the high costs associated with tooling and production, only 36 models were built. Muntz bought the company in 1951 for $200,000 and set out to improve the original design. Working with Kurtis and fellow designer Sam Hanks, the body was stretched 13 inches and made into a four-seater. Other advances included a padded dashboard on the passenger side, custom seats with belts, vibrant colors, and two cabinets — located below the armrests at either end of the rear seat. One of the armrests held liquor bottles and mixers; the other was an ice cabinet. Jim Muntz says his father sold a total of 320 Road Jets, each for $5,500, to friends in the entertainment industry, including actor Mickey Rooney, CBS CEO Frank Stanton, and Lash LaRue, a star in cowboy films.

Like his black-and-white portable television venture, Muntz shut down the car company, saying he lost around $1,000 on every sale. Fortunately, used-car sales were strong, and Muntz spent the latter half of the 1950s rebuilding his fortune.

## SOUND OFF

After King and Muntz parked their vehicles inside Ford's Wixom assembly complex, a small group of engineers listened to both systems. If any of the engineers noticed that King had installed higher-quality speakers, no one brought it up. To keep things fair, the same music was played on both units. After 20 minutes, King and his 8 Track system won the "Sound Off," and Muntz conceded the victory. One drawback was that Ford wanted to integrate tape players into dashboards, which the Stereo-Pak wasn't designed for. Before he left, Muntz asked King how Ford managed to keep rain from getting inside a convertible. It seemed Muntz had trouble keeping water from seeping through the edges of the Road Jet's removable fiberglass top. After King relayed that Ford used water-resistant rubber strips, Muntz offered King a bit of advice.

"He said artists like Frank Sinatra would never go for 8 Track because of the clicking (8 Track cartridges produced a slight clicking sound each time the playback head switched from one stereo channel to the next)," King says. "But Sinatra and a lot of the other artists came on board pretty quickly. RCA was behind it, and Capitol and Columbia got going soon after. It was a new medium where everyone made money, so it had a very compelling economic formula."

A fierce competitor, Muntz did not take well to the introduction of the 8 Track player and cartridge system. Soon after the launch of the 8 Track, he had stickers attached to his 4 Track players that read: "Don't ever buy an 8 Track tape player." Ed Campbell, who served as president of Lear Jet Stereo in Detroit, was not amused. "It was totally inappropriate, but that's just the way he was," Campbell says. "I got into it with him in 1970, when we were both invited to speak at a reception in Japan to open a new consumer electronics manufacturing plant. He was way out of line there, too."

After moving Lear Jet Stereo from Detroit to Tucson, Ariz., in 1970 (to be closer to a new 8 Track tape player manufacturing plant in Nogales, Mexico), Campbell prepared for the trip to Japan. He drew a $1 million letter of credit from a business account, which he planned to present to Toshio Niimi, the owner and president of Maruwa Electrical and Chemical Co., at some point during the weekend ceremony. Niimi, a successful business owner, manufactured consumer products in Nagoya, owned various nightclubs, and oversaw an FM radio station. A member of the Japanese Imperial Marines during World War II, Niimi hid in caves for more than a year after the war ended. Campbell, in the U.S. Marines, was on the island of Tinian, located south of Japan, at the same time as Niimi. "We never did find him on the island, but we shared some good stories when we first met," Campbell says. "His office had a wall with floor-to-ceiling sliding doors. Behind the doors was a large Seiko watch collection. You were invited to take a watch, and one for your spouse. I thought that was a nice touch, and we got along just fine."

At the dedication ceremony on Friday evening, Campbell recalls that Muntz, who was the featured speaker since the plant was building his 4 Track tape player, devoted most of his speech to the supposed shortcomings of the 8 Track system. "I was to follow Muntz, and by the time I got to the podium, I was fuming," Campbell says. "Muntz did nothing but criticize and tear down Lear. He said anyone was crazy to buy 8 Track. It was a big crowd, and he had no business doing that. I was not going to let him tear down the good name of Bill Lear, so I didn't even look at my original remarks."

Campbell kept his off-the-cuff speech short, congratulating the host for his technical and production prowess. He also refuted everything Muntz had said. "Toward the end, I reached into my pocket and said, 'Lear Jet Stereo is so impressed with Mr. Niimi and his new factory that I have a letter of credit here for $1 million to produce our players. That's the confidence Mr. Lear has in Mr. Niimi, Maruwa, and the already outstanding introduction of the Lear 8 Track.' I also mentioned that the consumer electronics industry had sold many, many thousands more 8 Track tape players than Muntz had sold 4 Track players."

As Campbell headed back to his seat, Muntz rushed to the stage and proceeded to tear down the 8 Track tape player again. He added that he couldn't stay for the rest of the festivities and was heading back to the United States to put Lear Jet Stereo out of business. But the damage had already been done. "The crowd figured things out pretty quickly," Campbell says. "Muntz had nothing to back up his statements, which were really his opinions."

After Muntz left, Niimi and Campbell cemented a deal over the weekend and formed Maruwa-Lear Jet Ltd. in Japan and Lear Jet-Maruwa Inc. in the U.S. The companies were successful for several years, producing thousands of 8 Track tape players before a series of deaths and accidents led to the closing of Maruwa. Following Niimi's death, his son took over, but he died after being struck by a car as he ran across Martin Luther King Drive during the Consumer Electronics Show at McCormick Place in Chicago. The banks chose his successor, but the new leader was involved in a serious car crash and died two months into the job. The next president died two months later from a heart attack. Following that, the company was closed for good.

Jim Muntz says Lear was adamant about producing his own cartridges. "There were some heated exchanges between my dad and Bill, and there was a lot of cash floating around, but my dad didn't have a bad bone in his body. He didn't want to kill the golden goose. Dad was committed to the royalty, and he couldn't get out of it, and that's why we attacked 8 Track."

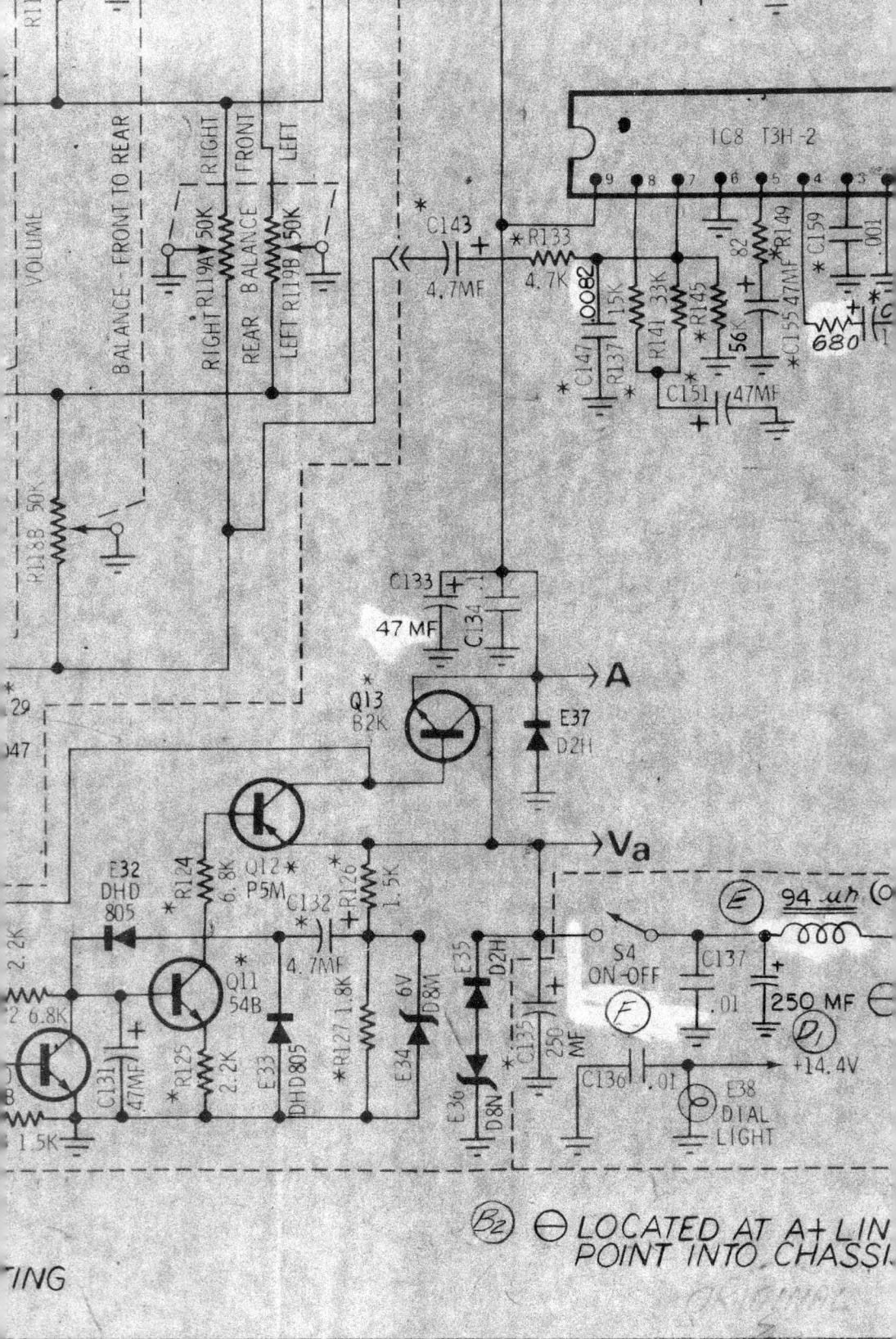

# MARCH OF THE ELEPHANT

## PACKARD VS. FORD

BORN IN 1878, THOMAS J. KING ARRIVED in Boston in 1899, having left Ireland for better prospects. He worked for several years unloading freight cars for B&O Railroad (Baltimore and Ohio). It was steady but uninspiring work, and he was eventually promoted to foreman. He did enjoy one claim to fame: A boyhood friend of Maurice Walsh, author of *The Quiet Man*, the pair would play rugby after school. As a measure of their close friendship, the lead character in two of Walsh's books, *The Key Above the Door* and *While Rivers Run*, was named after King. In 1952, *The Quiet Man* was made into a movie starring John Wayne and Maureen O'Hara, and directed by John Ford. The film was nominated for seven Academy Awards, and won Oscars for Best Director and Best Cinematography.

In 1906 King and his wife, Mary, moved to Chicago, then one of America's most prosperous cities due to its strategic position connecting the East and West coasts via rail. There, Mary gave birth to Mary Agnes King, the first of six children (four girls, two boys). An excellent student, Mary Agnes earned a teacher's certificate at the University of Illinois in Urbana, graduating with honors. While teaching at John Marshall Grammar School on Chicago's west side in the late 1920s, she met John Charles King, a native of Limerick, Ireland, and one of 13 children born to Edward and Ellen King. Like many children from large families of the day, John Charles, who was born in 1903 and emigrated from Ireland at a young age, graduated from high school but couldn't afford to attend college. A diligent worker, he rose through the ranks of the Chicago Surface Lines streetcar system, a forerunner of the Chicago Transit Authority. He married Mary Agnes King in 1934 (although they shared a last name, they were unrelated). The couple had four children: John Patrick, Barbara, Peter, and Edward. Unfortunately, John Charles caught pneumonia and passed away in 1940. Mary Agnes never remarried.

Growing up, John Patrick King's first job was delivering papers for *The Chicago Herald-American*, one of four daily newspapers in the Windy City in 1947. Twelve years old at the time, King had roughly 70 customers, and the route was near his mother's home. The afternoon paper cost 5 cents, and King made double the money over his competitors who hawked newspapers on street corners. He was a patrol leader with the Boy Scouts, and his hobbies included woodworking as well as learning about electrical and mechanical devices.

His first brush with a consumer electronic device came via his first employer. From a subscription contest, King won a Revere 8mm camera. Joined by a few other carriers, the small group was invited to Revere Camera in Chicago to collect their prizes. Following a tour of the manufacturing plant and the executive offices, actress Betty Hutton, who was married to Ted Briskin, a principal of the company, presented the cameras. Hutton starred in several movies, the most famous of which was *Annie Get Your Gun*, produced by MGM in 1950.

"I remember being a little nervous about meeting Betty Hutton, but I loved that camera," King recalls. "My friends and I would make movies, and we had to rent a projector to watch them."

Following the paper route, at age 14 King began working as a stock boy at a warehouse operated by Sears Roebuck & Co., at the time the largest retailer in the world. Customers would arrive with a receipt, and King would walk back into the warehouse to retrieve the item(s). He also worked in the mailroom at Keeney Publishing Co., which produced manuals for heating and cooling equipment, and he delivered telegrams for Western Union. Since there were no computers or fax machines, King would deliver the telegrams — the messages were printed on a quarter-inch-wide strip and pasted on a larger sheet of paper, and collect a signature so that Western Union had a record of the delivery.

King attended Our Lady of Sorrows Elementary School and St. Phillip High School, and proved to be a diligent student with a keen interest in math and engineering. In the summer following his graduation from high school — he was among the upper 10 percent of his class of 300 students — King found work at Admiral Corp., where he packaged radar sets destined for the United States military. Following that, King took another job for more money at Ryerson Steel. There, he helped guide crane operators as they moved large steel coils to different areas of the plant, or placed them atop large truck beds. "I'd also wrap thick chains around the steel coils and wave to the crane operator that they were secure," King says. "It was hot in the plant, but the pay was good." At the end of the summer, in 1953, King quit and enrolled at the Illinois Institute of Technology.

He didn't waste much time declaring a major, and pursued a Bachelor of Science degree in electrical engineering. Living at home, he commuted to school by bus or the elevated train for the first year, before becoming the principal driver for the family. The car was a spacious, medium green 1951 Packard. During his sophomore year he took a factory job at White Cap Inc., which had developed numerous advances in food storage — including a vacuum seal for food jars consisting of a plastic molded seal lodged inside a metal cap — since its founding in 1926.

The arrival of World War II brought on a greater dependency on glass jars in American grocery stores, as metal was rationed. At the plant on Central Avenue, which had around 300 workers, King was a pressman and a troubleshooter. If a machine or line broke down, King would fix it. Working the night shift, from 11 p.m. to 7 a.m., he took evening classes and slept during the day. His combined school and work schedule was such that he completed his degree in 1960.

By that time, King had been married for three years; he met Barbara Ann Bostrom at a mutual friend's Sweet 16 birthday party. Most everyone at Irene Whetter's party knew one other, and for a time the group would go out together. At some point, King asked Bostrom on a date. "Then it just became the two of us, and we were married on January 26, 1957," Barbara recalls. "We got an apartment in a large building in Chicago. It had a bedroom, a good-sized living room, a kitchen, and a full bathroom. It was neat and tidy, but we soon outgrew it."

The apartment was located on the first floor, although it was set some eight feet above the sidewalk. Like many dual-income families, the Kings only saw one another on weekends. "I was working as a beautician, and he would come home to sleep for two hours during the day," Barbara says. "To help him rest, I would take the two girls (Kathy and Linda) on long walks while he slept."

While he was dating Barbara, King and a friend, Frank May, made a wager. May, who owned a 1937 Ford Business Coupe, was confident he could drive a four-mile trip on the city's west side, from downtown to Garfield Park, using Lake Street, faster than King could make the same journey in his family's Packard. King chose Washington Boulevard, which was a quarter of a mile south of Lake Street. The day of the race, May and King positioned their cars on the respective streets and took off at the appointed time. During the race, May blew through various red lights. When he reached Garfield Park, May saw King standing next to the Packard, grinning wildly. "What Frank didn't know was that the lights on Washington were timed, which I think were set at 30 mph," King says. "I don't remember what the bet was for, but we remained good friends."

May's coupe came in handy. Cruising the streets of Chicago, six or eight friends would pile into the two-seater — two in the front, two on a large package tray set inside the rear window, and four in the trunk. "If we saw the police, we would pull the trunk lid down so they wouldn't see us," King says. "If we really worked at it, we could get five or six people in the trunk. It was just a huge space because it had been designed to accommodate traveling salesmen. We had a lot of fun in those days; it was just an innocent time."

## FAMILY TIME

In 1959, one year before his college graduation, King took a job designing various automotive aftermarket parts in the research lab at Shurhit Products Inc. in Waukegan, Ill., located a few miles north of Chicago. Shurhit, eventually acquired by Borg Warner Corp., produced distributor caps, rotors, ignition parts, and other components.

The added income from Shurhit provided the next step to middle class stability: the couple's first home, a white tri-level in one of Waukegan's middle class neighborhoods. Situated along the shore of Lake Michigan, the city initially developed as a shipping town before the economy became more diversified. Away from the hustle and bustle of Chicago, the new home offered three bedrooms and a full bath on the upper level, a kitchen, an eating area, and a living room on the ground floor, while a large rec room took up most of the lower level.

Like many families, the Kings had their share of memorable stories that served to entertain the generations that followed. Soon after the move, the circus pulled into town. After the big show was set up, and with the performers and animals easing into their routine, one of the elephants broke free from its bindings. It crashed through a perimeter fence and headed for the nearest neighborhood, located a few blocks to the southwest. When the King family arrived home from an afternoon excursion, they found a few of the redwood boards from their backyard fence scattered across the lawn. Several large footprints also were visible on the lawn. "We never saw the elephant, but we heard from the neighbors what had happened," King says.

After two years at Shurhit, King landed at Webcor, forerunner to Webster-Chicago Corp. Since its founding in 1914, the company had designed and produced state-of-the-art phonographs, tape recorders, public address systems, record changers for use with home Hi-Fi systems (High Fidelity), radios, amplifiers, and intercoms. In 1952, the company shortened its name to Webcor. One of the company's best-selling products was the tape recorder 210 Model. All previous reel-to-reel tape machines had to be flipped manually; a less-than-convenient prospect. With the flip of a lever, the 210 Model could play magnetic tape forward and backward. Webcor also pioneered a multi-selection TV knob, which proved to be a hit with consumers in the 1950s. Titus Haffa, who owned Webcor and served as chairman, operated a few related companies at the time, including Dormeyer, a producer of kitchen appliances, and Haber Manufacturing, which made motors for use inside mixers, toasters, coffee makers, and tape recorders, along with a record-changer that held 10 discs stacked on top of one another; an industry first. The player could operate at three speeds: 78, 45, or 33 1/3 RPMs.

"I worked on a lot of things at Webcor, including a pocket-sized dictation recorder called the Personic," says King, who specialized in solid-state circuitry and magnetic recording. "That's where I picked up most of my experience in tape recording, which was becoming more popular. The record companies were very strong. RCA had developed the 45, while the 33 1/3 came from Columbia. There was a lot of competitiveness, if not animosity, between the record companies in those days. They were looking for any advantage in the marketplace."

The Kings settled into suburban life, with a family of six children now in the house: Kathy, Linda, Nancy, Patrick, Robert, and Mary. Because it was so crowded, the baby slept in a crib in the master bedroom. Her three sisters occupied another bedroom, outfitted with two bunk beds, while the two boys held fort in the other.

Seven months after Mary arrived, a quarter-page ad in *The Chicago Tribune* caught King's eye. It was the fall of 1964, and Ford was looking for someone with experience in recording tape. "Beyond that, it was anyone's guess what the job entailed," King says. "I imagined that was just the way Ford wanted it. They didn't want to tip off General Motors or Chrysler that they were looking to introduce an 8 Track tape player." Soon after submitting his resume, King received a call inviting him to interview at Ford's research campus near the company's headquarters in Dearborn, Mich.

## PRODUCT PLANNERS RULE

In early 1964, the U.S. economy was still feeling the effects of a lingering economic recession. Economic policy missteps, high unemployment, and excess production capacity were the norm. Contributing to the problem was the fact that any increase in demand for factory goods was quickly met. Like its domestic rivals, Ford was eager for an economic expansion.

It came later in the year when the Kennedy-Johnson tax cuts took effect. At the time, the country was ramping up factory orders of military parts and equipment as tensions intensified in southeast Asia. When demand for labor soared, excess production capacity faded away. As the economy began a slow rebound, the product planners at Ford became bullish. The introduction of the 1964 1/2 Ford Mustang, developed in secret and introduced in April at the New York World's Fair, couldn't have arrived at a better time. Still wounded from the poor reception the Ford Edsel received in the late 1950s, Henry Ford II, chairman and CEO, wasn't eager to gamble on a new model. In fact, Ford II rejected the Mustang four times.

In the early 1960s, even diehard fans were unimpressed with the company's product offerings. "I clearly remember sitting around the dining room table and my kids saying, 'Dad, your cars stink. They're terrible. There's no pizzazz.' That started the whole thing," said Donald N. Frey, Ford's chief engineer.

To break through the mediocrity, Frey (pronounced Fry) assembled a team of trusted colleagues including Lee Iacocca, vice president of the car and truck group, and leading stylists John Najjar and Philip T. Clark. Raiding the company's existing parts supplies, Frey directed that the Mustang project be hidden away in a design studio. Iacocca came up with the idea of holding a competition for the project. In the end, the design team at the company's Lincoln-Mercury division, including Joe Oros and L. David Ash, among others, won out. Some reports say the Mustang was developed for a tenth of the cost of a new model. "The whole project was bootlegged, there was no official approval of this thing," Frey told *USA Today* in 1994, as part of the Mustang's 30th anniversary. "We had to do it on a shoestring."

Frey, who spent 18 years with the automaker and reached the level of vice president and general manger of the Ford division — he left in 1968 following a dispute with Iacocca — was a major proponent of the 8 Track tape player. With the success of the Mustang, Ford's powerful product planners pushed for more new models and features. Wedged between the company's top brass and various departments like engineering and radio, the product planners had a receptive ear when Bill Lear unveiled his 8 Track tape player.

Many product planners were prone to favor flash over substance. "Bill Lear was insistent that he run the whole show and supply us with his 8 Track players, but we wanted Motorola to build them," says Tom Walsh, manager of Ford's heating, air conditioning, and radio department. "We felt Motorola was better prepared to ramp up the volumes we needed. To be honest, Lear went around us and tried to cram the thing down our throats. The project was done too fast, in my opinion. The product planners were full of themselves, and while some of them may have had engineering degrees, they had no engineering experience."

Taking into account that a crash program could go sideways in a hurry, Frey green-lighted the 8 Track tape player in October 1964 (integrated with an AM radio either inset into the dashboard or hung just below it). Ford's normal product development cycle for such a player was two to three years, Walsh says.

It was a joint launch. Motorola supplied the players and RCA the music (using Lear cartridges). At the start, the combination AM radio and 8 Track player integrated into the dashboard was available as a $270 option in the Lincoln Continental, Ford Thunderbird, and Ford Mustang. A player mounted below the dashboard was available as a $140 option in the large Fords like the LTD and Galaxie, as well as comparable Mercury vehicles and station wagons. In the first year, 125,000 players were ordered from factories, while thousands more were sold and installed by dealers. Outpacing air conditioning, the 8 Track player was among the most popular options in Ford's history up to that time.

"It made quite an impression when you (were) on the line at the factory watching the first players installed," King says. "When I came into Ford (as product design engineer, radio and stereo), I reported to supervisor Fred Bauer. My unofficial title was project engineer for the 8 Track tape player. It was not top secret within the auto field. Lear was talking to other OEMs, and Ford was the one that elected to go with it. It was very positive. I do recall meeting with product planning periodically and telling them what we were going to do. If we had a technical or supply problem, product planning would ask what we were doing to solve the problem, so we would tell them what we were doing next. We would get things fixed very quickly."

# CARTS, DISCS, AND MUNTZ

## A BLOODLESS WAR

THE CITY OF TOLEDO WAS IN DIRE STRAITS when the Miami and Erie Canal opened in 1843. It had lost the long political debate over the location of the northern port that would connect the southern Ohio basin and its rich agricultural output with Lake Erie and the trading towns of Detroit, Cleveland, and Buffalo. Toledo's nearby rival to the north, Manhattan, had won out, and was poised for an economic windfall from the boats that were arriving laden with tools, furniture, farm equipment, and agricultural goods. The only concession Toledo gained from the debate was the addition of a sidecut that would give it access to the canal. Still, the many years it took to complete the project served to strengthen Toledo's competitive position. Canal planners hadn't accounted for the construction of ever-larger ships to transport goods. As bigger boats arrived, the crews soon favored Toledo's deeper ports on the Maumee River. Manhattan, for all its promise, was appropriated over the following decades and is now a footnote in history.

When George Eash came of age in the 1940s, Toledo was a bustling industrial town teeming with more than 300,000 people. Following World War II and the production of the thousands of Willys Jeeps used in the European and Pacific theaters, Chrysler and General Motors, along with numerous suppliers, converted their Toledo operations to meet civilian demand. While the Miami and Erie Canal was all but abandoned by the early 1900s, it had served to spark economic prosperity. In a profound way, the canal and others like it ushered America into the modern age — the great migration of people from agrarian villages to industries springing up in and around urban cities connected to large ports and railroads.

Attracted to a burgeoning electronics industry, George Eash, an engineer, had built up a successful audio and visual products company in Toledo, but he lacked the capital to expand the operation outside of the Midwest. In 1954, the business, Sound Electronics Laboratories, was sold to G.H. Paulsen, a successful insurance salesman in Toledo who quickly saw the potential for greater things. Now under the umbrella of G.H. Paulsen & Co., Eash, who took the title of chief engineer, was asked to develop a player that could accommodate 15 minutes of recorded tape, ostensibly so that Paulsen could play music in his car.

Reflecting on the request a decade later, Eash told *Billboard* that the idea for a closed-loop tape system wasn't a new one. In 1954, both Mohawk, a firm in New York, and Cousino, an electronics supplier in Toledo, were working on various tape projects. Freed from the task of raising capital, Eash set to work and soon crafted a plastic cartridge that worked with a rudimentary player. "I got 300 feet of tape going around pretty quick after some work, and so I knew it was practical to think of a continuous-loop cartridge system," he said. As Eash was showing off his system to RCA Victor — he calculated it would take 1,200 feet of tape to meet the record company's request for a cartridge that could fill an hour — his patron, Paulsen, became ill and died. The company went into receivership. Because Eash worked in relative obscurity, he took the prototypes with him.

Initially, the pipe-smoking Eash developed a system where the tape circulated inside a cartridge via a pinch wheel driven by a small motor; future models placed the pinch wheel inside the player. The latter setup didn't pose a problem since the player was envisioned for use in a controlled environment like a radio station or a doctor's office. Lacking the capital to manufacture prototypes, Eash partnered with Viking Co. in Minneapolis. After three patents were issued to Eash in 1957, Viking released the "35 Series" player, which retailed for $70. While the consumer market was slow to materialize, the player found a receptive audience among radio stations. As other companies jumped into the marketplace, cartridge makers found they could deliver a compelling economic advantage to radio producers. Rather than have engineers load cumbersome reel-to-reel players before every station break, a disc jockey simply pushed one of Eash's licensed cartridges into a machine and flipped a lever to activate the pinwheel. The length of the tape typically provided five minutes of playing time, enough to fit four or five commercials, a song, or a weather report. In 1956, Eash hired a Toledo advertising firm, which came up with the Fidelipac name.

By 1959, nearly every major radio station in America used closed-loop tape players. It couldn't have come at a better time. Five years earlier, the introduction of color television started a major shift in broadcast patterns. Programs such as "The Lone Ranger" and "The Green Hornet" got their start on radio, but after 1954, television, hungry for story lines, produced 30- and 60-minute shows around crime-fighters, police officers, and cowboys. To offset the competition for content, radio stations developed "Top 40" playlists where the most popular songs of the day were played back-to-back, interrupted every few minutes by a disc jockey.

As the technology progressed, more songs and commercials were made available on cartridges, which further reduced the cost to operate a radio station. It wouldn't be long before someone brought Paulsen's idea — a pre-recorded music cartridge and player integrated into a vehicle — to life. Not everyone saw the potential, however.

## HIGHWAY HI-FI

In 1915, dozens of automakers competed for America's purse strings. By the 1950s, GM, Ford, and Chrysler came to dominate the market, with smaller manufacturers like Packard, Studebaker, Nash, and Hudson holding their own. That changed in 1953 when Ford II, the grandson of founder Henry Ford, launched a price war in a bid to restore his grandfather's company as the leading automaker. Up until the late 1920s, Ford, along with its Lincoln division, was the dominant automaker. But GM, with its formation or acquisition of Chevrolet, Cadillac, Pontiac, Oldsmobile, and GMC, among others, took the top spot. Had Henry Ford moved faster to offer more vehicle lines beyond the Model T, the company would likely have entered 1930 as the market leader.

Following the Great Depression and World War II, Ford was near bankruptcy and barely had enough money to convert its factories back to automobile production. At the time, GM was the No. 1 automaker, followed by Chrysler, which topped

1 million in vehicle sales for the first time. Ford picked up the rear. By the early 1950s, the company was on its feet again, and Ford II, since named president, sent a tremor through the automotive marketplace. In 1953, he ordered drastic price cuts on nearly every model. GM and Chrysler, which rarely backed down from a fight, went toe to toe with Ford. The smaller players couldn't compete.

The following year, Nash merged with Hudson and was renamed American Motors Corp. George Romney, who in the next decade would become a three-time governor of Michigan, oversaw the operation. Following the Nash-Hudson merger, Packard acquired Studebaker (Packard ceased production in 1958, while Studebaker quit the market in 1966). Ford II, with his price war, set off one of the last major consolidations of the U.S. auto industry. American Motors, having acquired Jeep in 1970, eventually partnered with French automaker Renault before the whole enterprise was sold to Chrysler in 1987.

Two years after Ford II's price war opened up market share to the Big Three automakers, Chrysler was in an especially bullish mood. Around that time, Peter Carl Goldmark, an engineer at Columbia Records who was instrumental in developing the LP record in 1948, was confident a record player mounted under an instrument panel might interest an automotive company. Following a suggestion from his young son, who lamented that there were scant adventure stories on the radio, Goldmark began working on a mobile record player. In his book, *Maverick Inventor*, the man who developed the 33 1/3 phonograph says he ignored management and developed a record player for use inside vehicles in six months' time with "the narrowest groove in the business." He called it the "ultra microgroove."

Spinning at 16 2/3 RPM, the new, 7-inch records he and his team developed had three times the number of grooves of an LP. After testing several prototypes, Goldmark put in a call to P.J. Kent, Chrysler's chief electrical engineer. A week later, Goldmark drove from New York to Chrysler's headquarters and research campus in Highland Park, Mich., an island city within Detroit. Goldmark arrived in his specially equipped family Chrysler. Taking Kent out to the parking lot, Goldmark flipped a switch and a record player slowly descended from below the instrument panel. Next to the machine was a cavity to store six records.

"It's fine while you're parked," Kent said, after listening to the player, "but what about driving on the road?" Goldmark handed Kent his keys, and the pair took off down the neighboring Davison Freeway. The player passed the initial test, and Kent returned to the Chrysler campus and headed to the proving grounds. While the player was properly balanced, Goldmark hoped the weighted stylus could withstand the series of cobblestone streets, high-speed curves, small hills, and quick stops. Fortunately, the overall system held its own. A few days later, the tests were repeated.

Chrysler President Lester L. "Tex" Colbert slid behind the wheel and Kent took the passenger seat, while Goldmark and Chrysler Vice President Lynn Townsend got in the back. Much as the first day, the phonograph performed admirably. Returning to the executive garage, Townsend turned to Goldmark. "I must have it for Chrysler," he said. The other two men agreed. "Yes, we must have it."

The goal was to introduce the player in late 1955 and offer it as an option for the 1956 model year. One minor hiccup was quickly fixed, Columbia Records was on board to supply the entertainment, but Chrysler had to agree to purchase 20,000 players.

The project moved forward, seemingly successfully, until Goldmark received a call two weeks before the press launch. It seems the record player worked fine inside Chrysler vehicles, for which it was originally designed, but problems developed when the machine was installed in the automaker's sister nameplates. Goldmark caught the next flight to Detroit. "As soon as I arrived, the engineer put me in the car and started driving with the record player on," Goldmark said. "It was incredible. The machine wheezed, fluttered, groaned, jumped grooves, and made noises I had never heard before. It did everything it was designed not to do. What had happened? And then I glanced at the dashboard and almost jumped out of my skin."

Without consulting Goldmark, Chrysler's engineering team installed the players on Dodge and Plymouth models. "The characteristics of those cars are quite different from those of the Chrysler line," Goldmark wrote. "They were lighter and harder-riding, for one thing, with different kinds of suspension. Obviously a record player installed in these cars needed a different kind of dampening." Returning to CBS Laboratories, Goldmark and his team recalibrated the machines for the individual brands — Chrysler, Dodge, Plymouth, DeSoto, and Imperial.

On September 12, 1955, Chrysler issued a press release stating its intention to offer a record player "located in a shock-proof case mounted just below the center of the instrument panel." The automaker touted the fact that the player would not be affected "by the angle of the car, its highway speed, or even severe cornering." The so-called Highway Hi-Fi system was the first time a car owner could control what was played inside a vehicle, whether a Broadway musical, the music of Cole Porter, or dramatic programming. "And if the children are restless on a long ride, Davey Crockett and Gene Autry are ready at hand to help keep them quiet," the press release stated. Up to 45 minutes of music was offered on each side of a record.

Consumer demand proved to be tepid. Goldmark blamed lackluster marketing by Chrysler and Columbia. Content was another challenge. The Highway Hi-Fi system came with a boxed set of six records, but there was a limited selection of the 7-inch discs available in stores. What's more, dealers failed to get on board, the players were prone to break, the hardware attracted dust and road sediment, and drivers didn't much care for changing records and setting a needle while on the go. With warranty costs climbing, Chrysler pulled back on the record players in 1957 and discontinued them the following year. The poor results were not lost on GM and Ford. The in-vehicle playback market was set to the side. While consumers can be adventurous, it was quite a stretch to ask a driver to change a record at 70 mph.

## RISE OF THE TAPE

In 1946, Earl Muntz checked into the Warwick Hotel on 54th Street in New York City. Located within sight of the Empire State Building and its large antenna array, Muntz got on the phone and ordered three televisions — a Dumont, a Philco, and an RCA. When the sets arrived, Muntz turned the units on and set the channels to the same station (there were four stations at the time). Carefully, he disassembled the units and began pulling out tubes and other components until the picture disappeared.

His reasoning was straightforward. At the end of the experiment, he had a good idea of the exact components needed to operate a television set. As with many consumer electronics, engineers can "overdesign" products to ensure reliability and longevity. Muntz took the opposite approach — eliminate everything except the components needed to make a product work.

Returning to Van Nuys, the used-car salesman opened a research laboratory and directed his team of engineers to design a series of TV sets, including a portable device that would retail for less than $100. Fond of visiting the lab, Muntz had a habit of carrying around a metal clipper in his shirt pocket. When presented with a prototype, Muntz would turn on the unit and study it for a while. Soon after, he would take out his metal clipper and start removing small transistors and other components. If the device kept working, he would clip off more parts. When the unit failed to work, Muntz would have the last component reinstalled. He called his act of elimination "Muntzing," and used the phrase in advertising spots.

In 1948, the used-car salesman was producing 5,000 miniature black-and-white sets a month. To keep prices down, the units were engineered to work in large cities. He didn't bother to serve rural markets because the added components needed to pick up far-off TV signals were too expensive, according to the book, *Free Enterprise Land*. Soon, Muntz opened his own stores to sell TV models his team had developed, along with other electronic equipment. A master salesman, Muntz would mail out hundreds of television knobs at a time. Each one came with a note, "Call and we'll show up with the rest of the set." At the height of his consumer electronics business, in 1952, Muntz had 72 stores that collectively produced $55 million in revenue. Two years later, the industry was changing and the self-taught engineer strived to take advantage of the next wave of broadcasting.

Because a color television required many more components than a black-and-white set, Muntz and his team struggled to develop a new line. In addition, the TV networks weren't keen on licensing their technology to a competitor so early into a launch. As his team dove into the development of an affordable color TV set in 1954, Muntz ended production of the Muntz Road Jet. By 1956, the TV company was $5 million in debt and the value of its stock dropped from $6 million to $200,000. The following year, the enterprise declared bankruptcy. The business was eventually sold off and continued on without Muntz for a time.

Falling back on his used-car business, the self-taught engineer kept an eye on what was happening and eventually became aware of the Fidelipac cartridge developed by Eash. After meeting at a consumer electronics show in Chicago, Muntz convinced Eash to join him in Los Angeles. In 1961, the pair tried to interest Ford's Lincoln division in offering their 4 Track player in its vehicles. The automaker declined, reasoning the tape motor couldn't run at an even speed in extreme weather conditions. A year later, after selling Muntz Auto Stereo, Muntz reassembled his engineering team and, with Eash, introduced the Stereo-Pak 4 Track tape player. After signing local bands in and around Los Angeles, Muntz convinced Columbia to license some of its catalog for the Stereo-Pak in 1964.

With Eash's engineering skills, the length of the tape was extended to 40 minutes. From a Ford referral, Muntz and Eash tapped Barber Coleman in Rockford, Ill., for a motor that could run off a 12-volt car battery. "Barber Coleman must have supplied us with about 30,000 motors for the units built in this country before we started importing the motors from Japan," Muntz told *Billboard*. "In 1962, we sent (an early) unit over to Japan, and Clarion Manufacturing worked on them for about a year before they came up with a satisfactory unit. In the meantime, we continued building them here."

Strictly an aftermarket offering, the player was initially assembled utilizing quarter-inch magnetic tape. Eash and others developed a playback head that could read four tracks of music (two sets of stereo channels). In a simplified way, a stereo channel was set at the top of the tape, the other at the bottom. To move the playback head between the two stereo channels, Eash set a manual lever on the side of the unit. When one side of the tape was finished, the playback head was notched downward via the manual lever and the second round of music began to play. The tape was guided across the head via a belt-driven capstan and a motorized pinwheel. The arrangement worked fine, but the automakers were sour on it. They wanted a unit that operated seamlessly. Taking a lesson from Chrysler's Highway Hi-Fi system, they didn't want anything that would take a driver's eyes off the road.

Resolved to offer the lowest price, in 1963 the Stereo-Pak and its manual track changer was off and running. Muntz sold 1,300 units that year, and in 1964, he racked up sales of 18,000 players, retailing for $225 apiece. "The product took off like gangbusters and most of it was by word of mouth," Muntz told *Billboard*. "We didn't have to advertise too much. When we got under $99, we hit the kid market and one told 10 and that was that. They sold."

A Motorola AM Radio/8 Track tape player offered in the 1966 Ford Thunderbird.

RCA Victor 8 Track tape cartridge, "Andre Previn Plays Music of the Young Hollywood Composers," released in 1965.

Lear Jet Corp. headquarters in Witchita, Kansas, where the 8 Track tape player was developed in Summer/Fall 1964.

Bill Lear, left, in white collared short with short sleeves, inside the lunch room at Lear Jet Corp. with his development team.

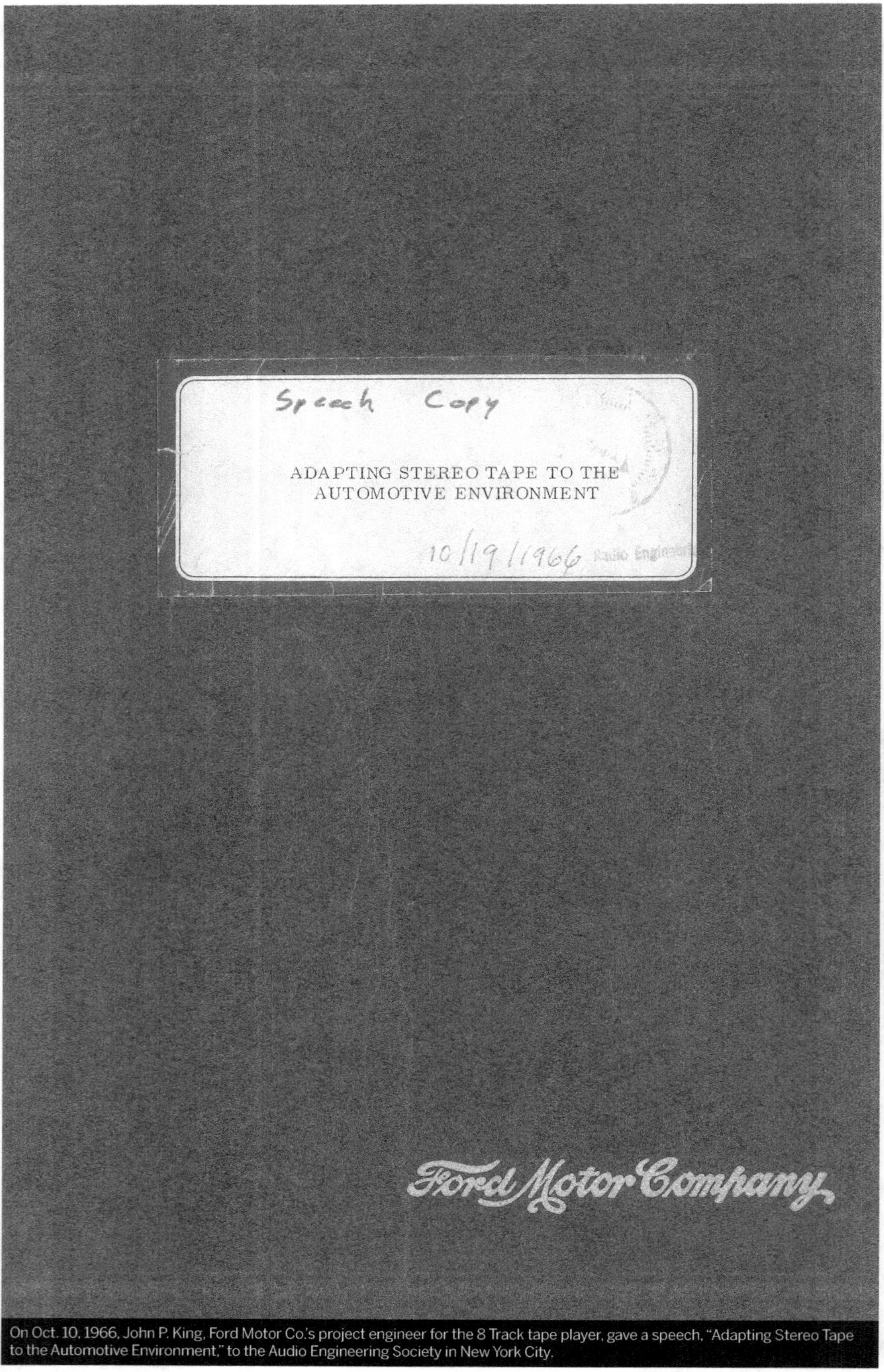

Speech Copy

ADAPTING STEREO TAPE TO THE
AUTOMOTIVE ENVIRONMENT

10/19/1966 Radio Engineeri

Ford Motor Company

On Oct. 10, 1966, John P. King, Ford Motor Co.'s project engineer for the 8 Track tape player, gave a speech, "Adapting Stereo Tape to the Automotive Environment," to the Audio Engineering Society in New York City.

*Good Afternoon: Gentlemen*
*I'ts a real pleasure to be*
*with you today.*

Radio Engineering

ADAPTING STEREO TAPE TO THE
AUTOMOTIVE ENVIRONMENT

Presented by:

John P. King
Radio Engineering Department
Ford Motor Company

to:

Audio Engineering Society
New York, New York
October 10, 1966

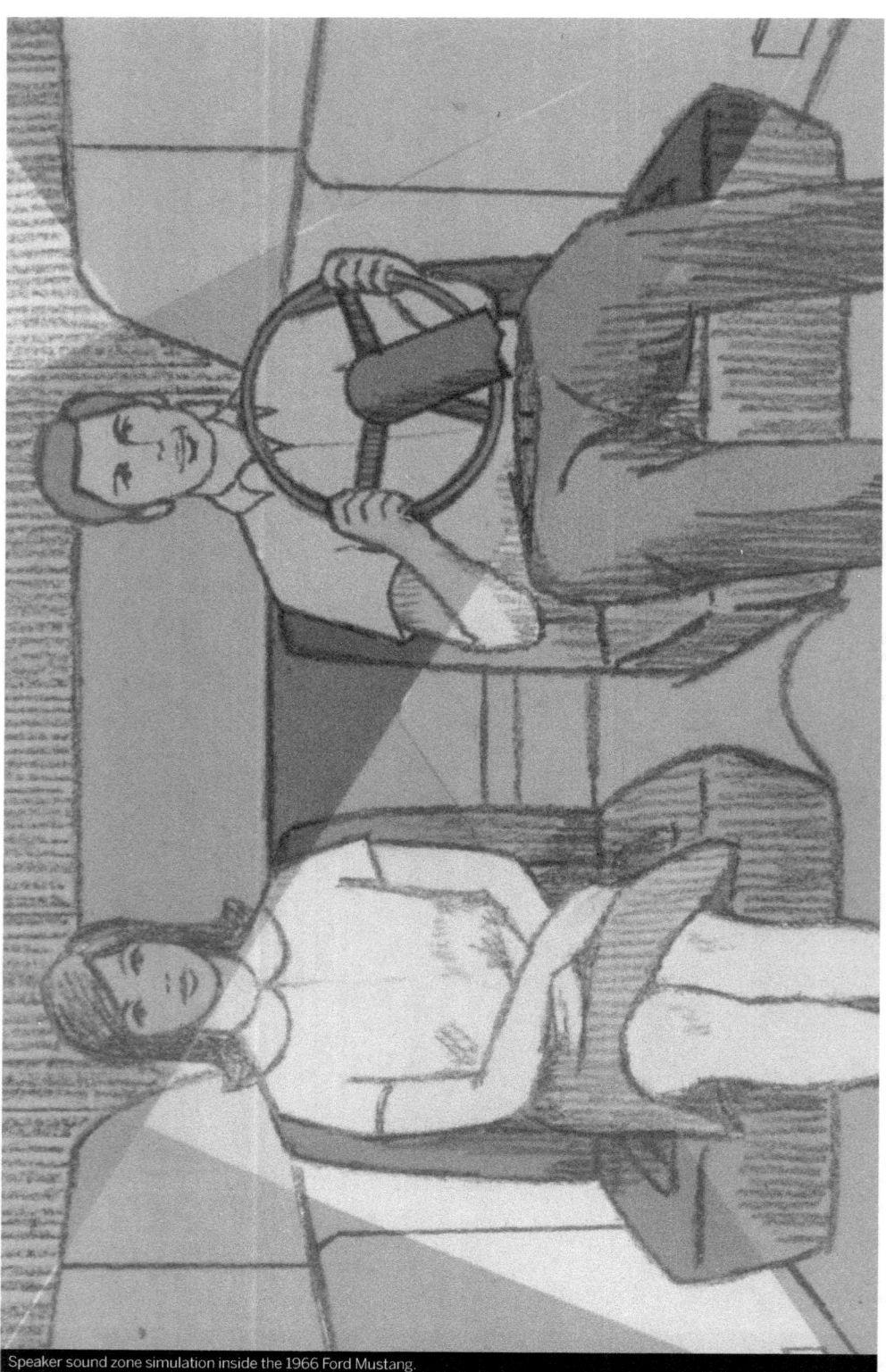
Speaker sound zone simulation inside the 1966 Ford Mustang.

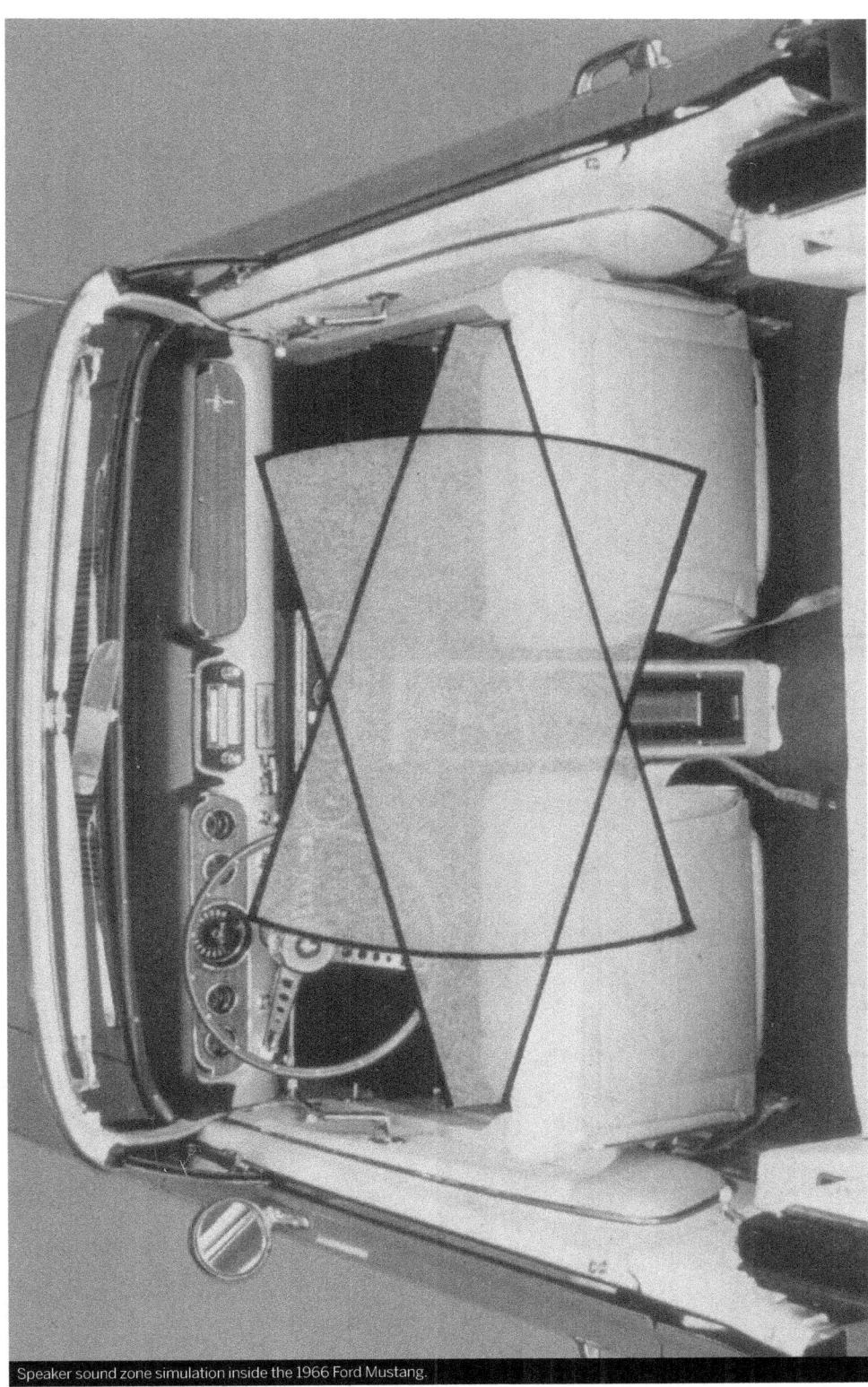

Speaker sound zone simulation inside the 1966 Ford Mustang.

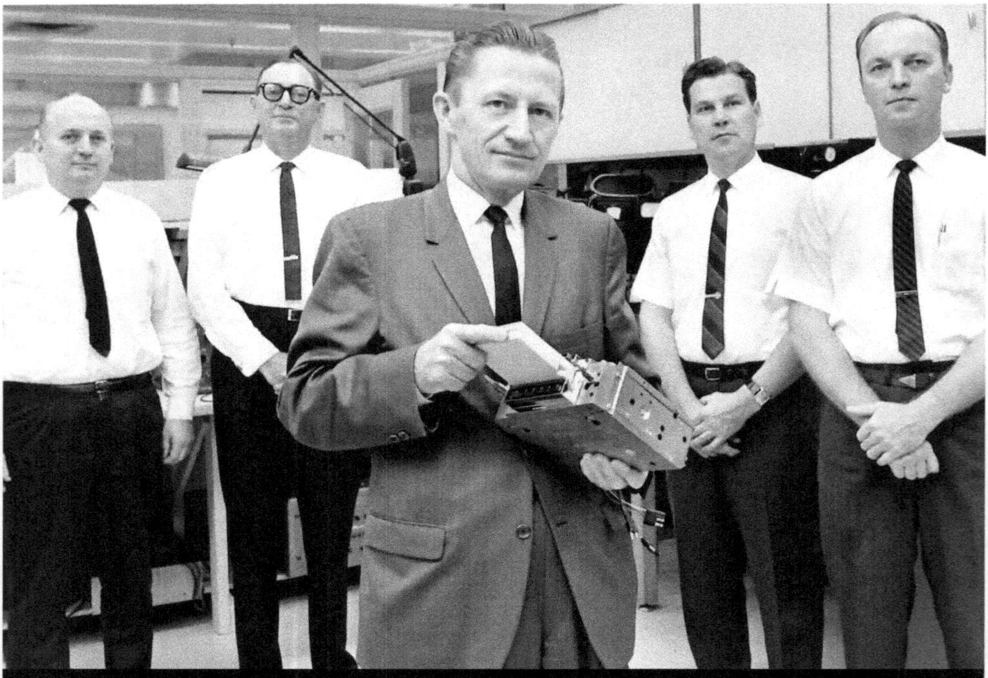
Oscar Kusisto, vice president and general manager of Motorola's automotive division, with his team at the corporate headquarters in Franklin Park, Illinois, circa 1965.

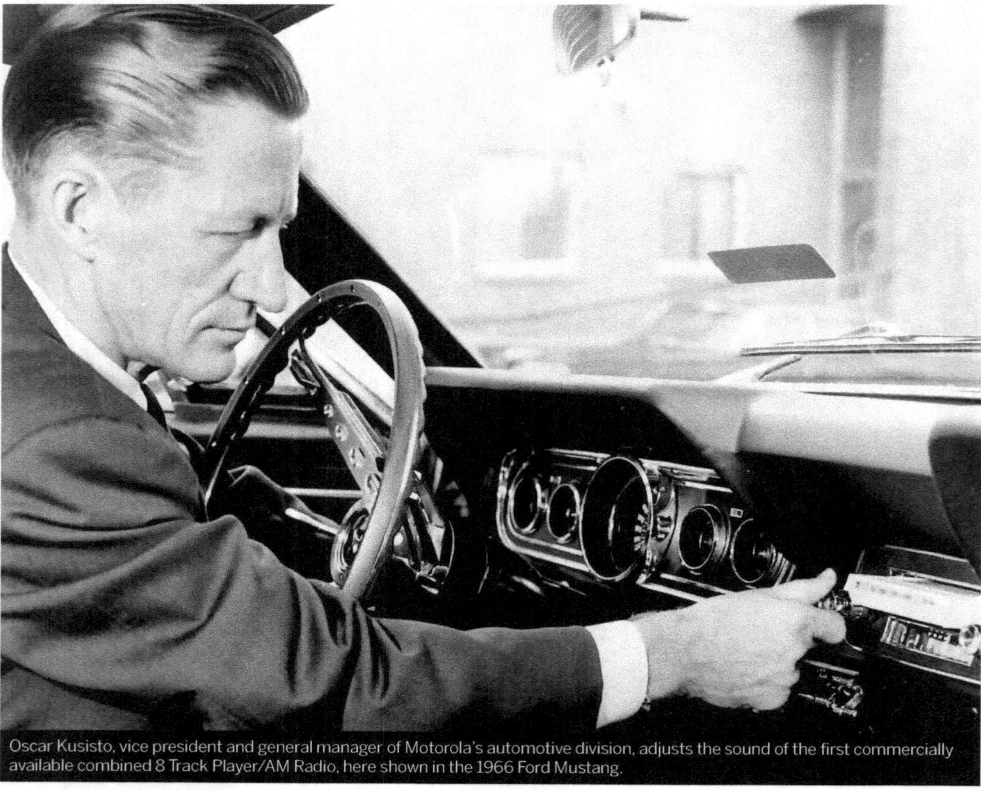
Oscar Kusisto, vice president and general manager of Motorola's automotive division, adjusts the sound of the first commercially available combined 8 Track Player/AM Radio, here shown in the 1966 Ford Mustang.

COURTESY OF KUSISTO FAMILY COLLECTION

Oscar Kusisto, vice president and general manager of Motorola's automotive division, with an 8-Track tape player for the 1966 Ford Mustang, stands in front of Motorola's headquarters in Franklin Park, Ill. (west of Chicago)

MANAGEMENT —
PRODUCT DEVELOPMENT —
MARKETING —

CONSULTANT: ELECTRONIC AND AUTOMOTIVE INDUSTRIES
DOMESTIC AND INTERNATIONAL

# OSCAR P. KUSISTO

570 Torwood Lane
Los Altos, California 94022

Phone:
(415) 941-9161

Later in life, Kusisto was a sought-after consultant.

Early tape heads.

A Ford engineer tests the early 8 Track tape player in a hot-cold environmental test chamber.

ADAPTING STEREO TAPE TO THE AUTOMOTIVE ENVIRONMENT

John P. King
Ford Motor Company
Radio Engineering Department
Dearborn, Michigan

Mass production of 8-track continuous loop tape
players is dependent upon parallel automotive
developments. The automotive environment with
its conditions of temperature, vibration, humidity,
and reduced space is much more hostile to tape
equipment than the more usual home application.
Unique contributions in the form of design speci-
fications, test methods, packaging, and speaker
developments have been necessary to make this
new automotive product commercially acceptable.

## INTRODUCTION

In the summer of 1965, stereo tape systems were installed in production

automobiles for the first time. This 1966 stereo system utilized an 8-track,

continuous loop, 1/4 inch tape cartridge and a player in the automobile's

instrument panel, integrated with an AM radio (see Figure 1). Both the

8-track cartridge and its player were entirely new and were not compatible

with then existing aftermarket equipment. This system has since been adopted

by all U.S. automobile manufacturers and by several large recording companies.

This system operates by sliding the tape cartridge into a slot in the

player. This forces the cartridge integral pressure roller against the

capstan shaft of the player and operation is automatic. A stereo pick-up

head is used to readout two of the eight stereo tracks at a time. Indexing

of the head between the four stereo programs is accomplished by a solenoid-

actuated cam. This indexing is automatic, triggered by a short strip of

metallic foil on the tape, with each program following in sequence (see

Figure 2). Selection of any of the four programs may be made by pushing

a manual override button. The 8-track cartridges operate at 3-3/4 inches

per second and are available in single and twin pack versions, equivalent

to one and two L.P. records, respectively.

# GO
# FLY
# A
# KITE

## A BLIND DATE

RAISED NEAR GENEVA, SWITZERLAND, in the 1950s, where her father attempted to develop a corporate jet, Shanda Lear was accustomed to opening doors for herself. Following the end of World War II, Bill Lear was riding high. The commercial aircraft industry was still in its infancy, and nearly every plane built after 1940 was equipped with Lear-designed and patented guidance and electronics systems. Eager to break into the European market, Lear set up a manufacturing plant in Switzerland. In 1955, he became enamored with the P-16, a prototype Swiss-designed jet. Done in by poor electronics and other system failures, five planes were built before the Swiss government canceled the project. Still, Lear was impressed with the P-16's wing design, which offered a series of slats and flaps along the front and rear edges. Confident he could solve the electronic problems, Lear embraced and substantially modified the aerodynamic design — the aircraft could take off or land with just 1,000 feet of runway, a remarkable feat for the time. L.E. Liepold, in his book, *William Powell*

*Lear: Creative Designer and Inventor*, described the scene. "(Lear) employed the firm of engineers who had designed the P-16 to do the basic aerodynamic work on a jet plane which would be used essentially in industry as a business jet, and then went to work enthusiastically developing the plans."

Over time, Lear found the Swiss government slow and, at times, unresponsive. In 1962, he ordered the whole enterprise be packed up and shipped to Wichita, Kan. Set in south central Kansas, Wichita, the Air Capital of the World, was home to aircraft companies like Cessna and Beechcraft. Settling in, Lear and his team developed the Learjet and begin filling orders in the late summer of 1966.

With 110 patents issued over his lifetime, Lear was a wealthy man. In addition to a home in Wichita, he and his wife had purchased a mansion in Pacific Palisades, Calif., just north of Santa Monica. After the Lears moved to Switzerland, their California home was rented to movie star Grace Kelly, who resided there for a year before marrying Prince Rainier II of Monaco in 1956. In the summer of 1963, Shanda Lear was staying at the house in California. Invited on a blind date, the 19-year-old was startled when Jim Muntz pulled in front of the mansion in a limousine. "I was impressed, to say the least," she recalls, "but I almost didn't make it inside. In Switzerland, a lady learns to open the door for herself. So when I approached the limousine, I reached for the door, but Jim grabbed my arm so tight that he was hurting me. He said, 'A gentleman always opens a door for a lady.' I was steaming mad by this time, but I climbed into the car. He told me he would take me wherever I wanted to go. I didn't want to go anywhere, but I couldn't think of a gracious way to get out of the date."

A self-described gadget queen, Shanda noticed a console equipped with buttons and knobs in the rear compartment. Curious, she turned it on. As music filled the cabin, she learned from Jim that it was a Stereo-Pak player that had just been introduced by his father, Earl Muntz. Remembering that her host said he would take her anywhere, Shanda directed the vehicle to the Santa Monica Airport. Her father would soon be landing in a Learstar, a small passenger transport aircraft that the inventor had refashioned from the Lockheed Model 18 Lodestar. The aircraft was originally developed for use in World War II. "Dad got off the plane and I ran up and grabbed his hand and brought him to the limousine to see the device," she recalls. "Dad was quite impressed with it, and he asked Jim who made it. The next thing you know, we're at Earl Muntz' house and we wound up eating sandwiches in the kitchen and talking about the Stereo-Pak. Dad had a great time."

Soon after, Lear installed a Stereo-Pak unit in his car, but he quickly found several faults with it. He also signed on as a distributor in Wichita, ostensibly to offer the 4 Track players on his planes. "There was some jarring and vibrations in the player, and Bill very much did not like the lever on the side to switch between the tracks," Campbell says. "He tried to give Muntz suggestions on how to improve upon it, but, both men being strong-willed, Muntz wouldn't hear of it. One thing led to another and Muntz told Lear to go fly a kite. Lear said he would do Muntz one better and develop an 8 Track tape player. Well, Muntz yelled back that putting 8 tracks on a quarter-inch tape was impossible (seven tracks was thought to be the limit). So Bill brought the project to Wichita and pulled several engineers off the aircraft project."

## A PATENT FOR THREE

Frank Schmidt, who along with Sam Auld, was part of Lear's engineering team, agreed to work on a new player. Coming over from the jet project, Schmidt joined Lear as a project engineer in the summer of 1964, just as the player department was being reorganized. The team initially tried to refine Muntz' Stereo-Pak, but the motor proved inconsistent. What's more, the rubber drive belt between the capstan and the motorized flywheel was prone to slip, which could stretch the tape (as could a bump in the road). The setup "just didn't work," Schmidt said in a May 1999 interview with Malcolm Riviera at *8TrackHeaven. com*. "Too much wow and flutter, and (it had a) problem with speed control. It had a lot of problems."

Schmidt said the Wichita setup was unique. "It was the only aircraft plant I ever worked in that had a barbershop," he said. "Bill felt that your hair grew on company time, so it should be cut on company time. You could call down there, get an appointment, and get a hell of a nice haircut. The other thing we had was a kitchen. It was a walled-in area right in the middle of the building. You could go in there 24 hours a day and you'd find a nice big kitchen with four or five tables, and everything you'd find in a kitchen — stove, sink, refrigerator, freezer, oven, the whole works. Completely stocked. Dishes, food, anything you'd want. It was all free."

Abandoning the Muntz player, the design team pursued Lear's idea for an 8 Track tape device. Instead of a belt system, the group developed an integrated direct drive motor to rotate the capstan. Schmidt explained the motor was nicknamed the "pancake," since the flywheel was 4.7 inches in diameter. To offset road bumps and other fluctuations, the Lear team reversed things. "In a conventional motor the armature rotated, and the field was stationary," Schmidt said.

"In Bill's motor, the field rotated and the armature was still. Our group developed this thing, along with the first players. About 75 percent of my time, though, was spent developing the cartridge."

Mindful that the cartridge would compete with the variances of human activity, the plastic unit had to be sturdy enough to withstand sudden pressure. It would be sat on, stepped on, dropped from different heights, sandwiched between seat cushions, and pinched by a door or a glove compartment. In addition to outfitting the players in each of his planes, Lear favored developing the unit for automobiles as a way to complement the radio. Eager to jump in — the inventor often slept in 20- to 40-minute intervals followed by several hours of work — Lear never tired of trying new things. To make his point, he was fond of reciting the tale of two frogs trapped in a large jar of cream. Both frogs struggled mightily to get out, but after some time, one frog gave up and drowned. The other frog kept trying and, after a while, the cream turned to butter and he escaped.

Auld recalls Lear would come into the 8 Track lab and would be "running a lathe, handling a soldering iron, (and) making things in the middle of the night. Can you imagine the chairman of the board doing that?" he said in Victor Boesen's book, *They Said it Couldn't be Done: The Incredible Story of Bill Lear*. The inventor was adamant that the players operate seamlessly with no levers or buttons. The driver should never be distracted, he reasoned. After several weeks of working on a plastic cartridge, Auld and Schmidt claimed it could be thrown against the wall with no damage except to the wall. It was "the simplest tape transport system ever devised," Auld said. With the player and cartridge system under development, Lear put in a call to the leading suppliers of tape heads, Nortronics Co., based in Minneapolis, and Michigan Magnetics, which had its operations in Vermontville, Mich. He requested a head with a much narrower playback field so that eight tracks could be read across a quarter-inch tape, from bottom to top. By the early 1960s, closely stacked multitrack heads were in use in the recording and computing industries.

With a dedicated playback head in the works, Lear and his team turned to the lever. In Eash's design, a built-in pinch roller flipped up and created the pressure to move the tape through the cartridge and start the audio playback process. To rotate the playback head from one stereo program to the other, the user flipped a lever. Lear and others quickly determined the system couldn't be adapted to play eight tracks, or four stereo programs. Few people would be willing to flip a lever four times during playback to reach each stereo channel. To get around the problem, Lear and his team placed a pressure roller inside the cartridge, while the drive motor was set in the player. The tape was guided in between, resulting in a near-automatic system. Once a cartridge was pushed into a player, music streamed instantly. All that was left for the driver to do was adjust volume, tone, or track selection. It also was integrated with an AM radio. Lear "was interested in getting around the patent that Muntz had. And that's where the (cartridge) pinch roller came," Schmidt said. He added Lear didn't give too much credence to patents. "I remember a discussion once where (Lear) said, 'Find a way to design around it, or we'll go ahead and do it and fight it out in court later.' "

A side-by-side comparison of Muntz' 4 Track player and Lear's 8 Track player is revealing. As an aftermarket accessory, the Muntz unit was designed to fit under the dash easily. As a result, the player hung from a metal plate, leaving the region between the plate and the player exposed. With some of the components visible to the eye, it was easy to see the setup might injure the hand of a small child. It also tended to collect dust, dirt, and ashes. Aware of the drawbacks, Muntz shouldered on with his original design. He eventually offered a catalog of several thousand titles.

Lear's player was self-contained. Other than a light metal panel that flipped up when a tape was inserted, the unit had no other exposed areas. While a child could stick a hand past the metal panel, it wouldn't get very far. And the motorized capstan, or rotating spindle, didn't engage until a cartridge was fully inserted.

The next task presented itself in obvious fashion: The playback head had to automatically move down three times and then move back up to the original position (a total of four positions) to pick up each of the four sets of stereo tracks on the tape. Given that a manual control was out of the question, Lear, Auld, and Schmidt came up with a solution: The trio created a solenoid-activated cam that moved the playback head down and up on the magnetic tape by rotating the cam in 45-degree intervals. To signal the head to move, a short strip of aluminum foil was spliced on the magnetic tape. When the aluminum strip passed over the playback head, it sent a signal to the solenoid to move the head 45 degrees. A manual button would perform the same feat. "Yes, we were all told we were crazy to move the head, but it was easier than moving the tape," Schmidt said. "At this time, an 8 Track play head was thought to be impossible (given) the cost involved. The heads were very expensive and custom-made." If Lear could create a steady market for the tape heads, Nortronics and others would invest in assembly equipment to meet demand, and thereby lower costs. With the mechanics solved, and the patent applications filed, Lear set to work rounding up customers. The goal: Distribute the players and cartridges to a mass commercial market, even though it would take time away from his Learjet program, which wouldn't get off the ground for another two years.

## HANK THE DEUCE AND GENERAL SARNOFF

The start of any new industry is often thankless, complicated, and filled with doubt, especially among potential investors. Breaking away from a mature industry was even harder. But few startups have the star power of Bill Lear. The launch of the 8 Track industry normally would have taken many months to materialize. It took Lear two phone calls in the late summer of 1964. Lear had a portion of the four-part program, the players and the cartridges, figured out. What he needed was a mass-market customer and a library of music. Given Lear's history as an early developer of the car radio, he saw the automotive market as the best option to roll out the new medium. It didn't hurt that his Rolodex spanned numerous industries.

To get things started, Lear reached out to his longtime friend, Ford II. Since being named chairman and CEO of Ford Motor Co., Ford II held great sway. Unlike the CEOs of most other corporations, Ford II didn't have to wait for a proposal to be studied by a committee. After hearing Lear's pitch, Ford II was agreeable to the idea of introducing an 8 Track tape player. That was all Lear needed to hear. He took the OK as a directive and lit a fire under the automaker's powerful product planners. The group was divided into two camps: Those who were enamored with Lear, and those who were reluctant to call the chairman and a member of the Ford family to corroborate the story.

As a sign of their bond, Lear arranged to have one of Ford II's cars shipped to Wichita for a special installation. "I remember when they brought Ford II's personal black Lincoln to the plant in a semi," Schmidt told *8TrackHeaven.com.* "They shipped it down from Detroit. We took it into our model shop and proceeded to completely remove the entire dash. I remember it had a beautiful dash — die-cast aluminum with striping. We put it on a milling machine and cut the openings for one of our players for a custom-mounting unit. We ended up breaking the casting, so we had to order another one from our local Ford dealer. We (broke) three of them before we got one that worked. But it was a really neat installation."

In addition, Schmidt installed a player in Lear's Learstar. "Another one of the interesting car installations was in 1965. We showed up at work and everyone was gathered around a car ... it was Elvis Presley's gold Cadillac, (a gift) from RCA Records. We put two 8 Track players in it and about eight speakers." Asked why Presley needed two players, Schmidt responded: "I don't know. That's what they wanted. It was a beautiful car. Everything on it was gold-plated."

Next, Lear rang up David Sarnoff in New York, a longtime friend and a fellow pioneer in radio. Early in life, Sarnoff, a Russian immigrant, had been content to join the newspaper industry. His career path took a turn when he joined a cable company in New York as an office boy. When a supervisor refused to grant him an unpaid day off in 1906 for the Jewish holiday of Rosh Hashana, he left and joined Marconi Wireless Telegraph Co. of America. Over the next 15 years, Sarnoff developed a keen interest in radio, which at the time operated much like telegraphs and telephones — one person communicating directly with another. Sarnoff envisioned radio as a direct-to-mass-market system, where one person could reach multiple listeners. He wrote his superiors on several occasions and encouraged management to develop a "radio music box." The idea went nowhere.

The tide changed in 1921 when General Electric acquired Marconi and re-named the enterprise Radio Corp. of America, or RCA. Eager to show off the mass market potential of radio to the new management team, Sarnoff, in July 1921, helped arrange a live radio broadcast of a world-championship boxing match between Jack Dempsey and Georges Carpentier. An estimated 300,000 people listened to the bout, and soon after, demand for radio sets took off just as Sarnoff had predicted it would. As he moved up through the ranks, RCA acquired its first radio station in 1926, after which it launched National Broadcasting Co. (NBC) in a bid to establish a radio network throughout America. Three years later, under Sarnoff's direction, RCA purchased Victor Talking Machine Co., the largest producer of records and phonographs in the nation at the time. In 1930, he was named president of RCA (then referred to as RCA Victor). A few months later, following an antitrust suit, the company was broken up. With General Electric and fellow investor Westinghouse no longer in the picture, Sarnoff took control. In addition to building up programming for AM radio stations and acquiring RKO, a film production and distribution company, RCA funded and eventually succeeded in establishing NBC as a major player in the television industry.

In the early days of World War II, Sarnoff, serving on the communications staff of Gen. Dwight D. Eisenhower, expanded various radio circuits, including Radio Free Europe, a broadcasting system that eventually reached all of the Allied forces in the war theater. In gratitude, the U.S. Army awarded Sarnoff the brigadier general star in late 1945. From then on, he was referred to professionally as Gen. Sarnoff or, simply, The General.

"After hearing Bill's pitch for a music catalog to support the 8 Track, Sarnoff licensed 175 albums to it," says Ed Campbell, who joined Lear Jet Stereo in April 1967 after his employer, Gates Rubber Co. (now Gates Corp.), acquired the various Lear companies, including Lear Jet Aviation. A few months after the purchase, Lear, who was elected chairman of the new concern, directed Campbell to secure the rest of RCA's music library.

Arriving at the RCA Building at 30 Rockefeller Plaza, Campbell passed beneath the frieze at the main entrance that reads, "Wisdom and Knowledge shall be the stability of thy times." As the elevator headed to the 28th floor, Campbell says he half-hoped it would stop at the 6th floor, where "The Tonight Show Starring Johnny Carson" was produced (it was moved to Burbank, Calif., in May 1972). Arriving at his destination, he called on Irwin Tarr, RCA's division vice president of recording tape marketing. "I waited outside Tarr's office for hours, but he wouldn't grant me a meeting," Campbell recalls. "This went on for a few days. When I finally called Bill and told him Tarr wouldn't give me the time of day, Bill told me to sit tight. He'd fly into New York that afternoon and would get things straightened out. I didn't know what Bill could do other than burst into Tarr's office and demand the additional recordings."

Late in the afternoon, Lear and Campbell met in a hotel lobby and the pair proceeded to Sarnoff's apartment. Following dinner and drinks, it was decided Campbell would meet with Tarr the following morning. "The next day, a chauffeured limousine picked me up at my hotel and took me over to the RCA Building," Campbell says. "When I got to the 28th floor, I was promptly ushered into Tarr's office. No more waiting. I could tell he was pissed off, but there was nothing he could do. Long story short, he turned a great deal of the RCA library loose. No one crossed The General."

**ON DEMAND**
Push-knob tuning and
other features were
added as the 8 Track
player/AM radio evolved.

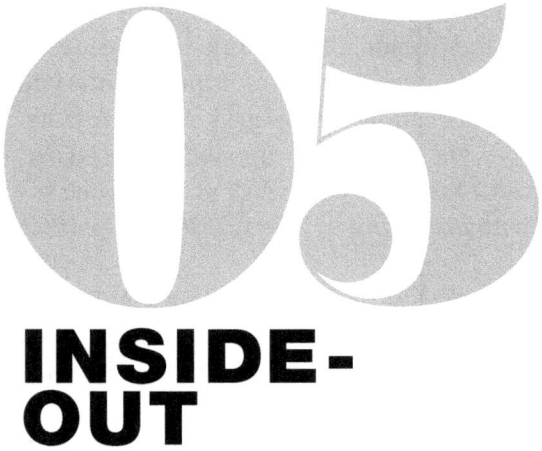

# INSIDE-OUT

## BACKSEAT DRIVER

BILL LEAR AND HIS ENGINEERING TEAM invented the 8 Track tape player and cartridge, but it was Oscar P. Kusisto, vice president and general manager of the automotive products division at Motorola Inc., who got the industry off the ground. Initially, Lear and his team produced 100 players and dozens of cartridges to generate interest and, consequently, purchase orders. As the first units were circulated in the automotive and home appliance industries in September 1964, Kusisto liked what he saw in the cartridge. It was light, durable, compact, and easy to use and store. Lear Jet Stereo's player was a whole different matter. Lear and his team over-engineered the player in a bid to increase stability. The development of the "inside-out motor" added considerable weight and cost to the unit. Kusisto, traveling between Motorola's headquarters in Franklin Park, Ill., located outside of Chicago, and a 40,000-square-foot office and production facility on Detroit's west side, convinced Motorola's management team that the industry would be better served with a lighter, less expensive player. Working through Motorola's global supply base, the motors were manufactured in Japan and shipped to the U.S. by air. Factoring in air freight, the motors were a third of the cost of those built in the U.S., and they were often more reliable. Most of the other parts were domestically sourced, including transistors, circuits, coils, springs, rollers, and diodes. Final assembly was completed at Motorola's sprawling radio production facility in Quincy, Ill.

Located along the Mississippi River on the far west side of the state, Quincy was named after John Quincy Adams, the sixth president of the United States. In the 1820s, the center of town was known as John Square, and Quincy became the seat of Adams County. For a time, the president's full name was accounted for. Later, John Square was renamed Washington Square. At its peak, in the late 1960s, the 990,000-square-foot Motorola complex employed 4,000 workers. Established in 1946 with 50 employees, the electronics supplier produced car, home, and portable radios at the outset. One of the company's more popular offerings, introduced in 1952, was a battery-powered "Pin-Up" clock radio that could be hung on a wall.

At the outset of the 8 Track program, Kusisto reviewed several options, including a compatible 4 Track and 8 Track player. Recognizing the Muntz 4 Track player had been firmly established in some regions of the country, Kusisto was open to the idea of a combined unit.

In October 1964, Kusisto received a directive from Frey, Ford's chief engineer. The automaker was determined to be the first to introduce an 8 Track tape system. Knowing Lear was approaching other automakers about the new technology, was Motorola prepared to deliver the first players the following summer to meet the 1966 model year launch? "Ford was so convinced of the ideal timing in the marketplace that the normal two-and-a-half to three-year procurement cycles were reduced to nine months," Kusisto wrote in an industry retrospective. "All traditional time schedules were breached. Long lead tools were ordered immediately, as design was completed. Three parallel engineering programs were launched: One program for current production models; a second for monitoring possible areas of concern as insurance; and a third for developing the next generation of advanced product. Through Herculean efforts, all obstacles were surmounted and production shipments began in July 1965."

Left unsaid was a drive by Kusisto to reduce costs. With Lear soon to be in possession of various patents for his 8 Track player and cartridge (which occurred the following year), Kusisto didn't believe Motorola or its partners should pay royalties. His reasoning: Motorola, Ford, Lear, and RCA were "cooperating" on the new player. What's more, Lear still enjoyed good relations with the top brass at Motorola, namely the Galvin family. Founded by Paul V. Galvin in 1928 with $565 and six employees, the Galvin Manufacturing Co. initially produced a device that eliminated the need for certain batteries in radios. Lear, for a period in the 1930s, owned a share of the company; the result of his first successful car radio program. After the company was renamed Motorola, Lear sold his share to help finance his advancements in aeronautics.

Following Motorola's lead, RCA and Ford took the same position on the Lear patents. "Bill Lear could be a bull in a China shop, but he wasn't going to undermine Oscar by making an appeal for royalties with upper management," Campbell says. "Motorola never paid a dime in royalties. Ford and RCA never paid anything. Everyone just wanted the player to come out, and there wasn't time to pick a fight. We did eventually get royalties from most everyone else, to the tune of about 25 cents a player."

Motorola also modified the player design. "In the course of development, Lear was developing its own player," King recalls. "Normally, you would follow the industry specs we had written. With our system, the tape head would start to read the top of the tape and work its way down to the other tracks before working its way back up to the top. For whatever reason, Lear decided to start the process from the bottom. So RCA, Ford, and Motorola got on Lear to conform to the industry standard, and they changed their player."

Kusisto and Frey, among others, had little confidence in Lear's ability to start up 8 Track tape player production from scratch. Given Motorola had been producing car and home radio sets for years, the supplier had the engineering talent and production capacity to produce combined AM stereo radios and 8 Track tape players for the automotive, residential, and marine markets. Early on in the program, much of the focus was on the Ford player.

Lear, all sides agreed, would produce parts and cartridges in Detroit and ship them to RCA's record pressing plant in Indianapolis or Motorola's factory in Quincy. To help speed the supply of cartridges, Motorola sold its office and factory on Detroit's west side to Lear and moved to an office facility on Michigan Avenue in Dearborn, which wasn't far from Ford's world headquarters, the so-called "Glass House." In addition to supplying the initial cartridges, Lear sought to license his player to other automotive and home appliance companies.

Keith Zerschling, a patent attorney at Ford from 1957 to 1991 (he retired as lead patent counsel), recalls visiting Lear in 1965 in Wichita to discuss a licensing agreement. "He was building jets at the time, and I believe he had four of them built," Zerschling recalls. "We had lunch close to his office, and we discussed the licensing arrangements. I don't recall Ford ever paying anything, but I know at the time there was bad blood between Lear and Motorola. It was eventually smoothed over, as I understand. Bill was to the point, and he was very much a fan of Ford. We just had to make sure everything was done correctly, and we protected Ford's interests."

## BUILDING 5

A package arrived at John King's home in Waukegan, Ill., in mid-October 1964. The large manila envelope included a letter inviting him to interview with Ford's heating, air conditioning, and radio department, along with a round-trip airline ticket. After arriving at the personnel department in Building 5, part of a grouping of engineering and testing facilities along Rotunda Boulevard just west of Ford's Dearborn Proving Grounds, King was escorted to an office at the east end of the third floor. There he met with Fred Bauer, the department supervisor. "Because I had a background in tape systems and solid state circuitry, Bauer revealed Ford was working on an 8 Track system with Motorola," King recalls. "I actually had interviewed with Motorola prior to that, but they never extended an offer."

Returning to Waukegan the same day, King went back to work at Webcor, where he was overseeing various tape recording projects. A month passed, and then another package arrived from Ford. Could King return to Dearborn and interview with Tom Walsh, the department manager? "I was hesitant to return to Dearborn because I had already been there, but I went after giving it some thought," King says. A few days later, a third package arrived at the house. It was an offer to join Ford as the project engineer of the still-secret 8 Track tape player program. "The offer was a little less than what I was making at Webcor, but the benefits were better," he says. "Plus, I thought the 8 Track had a lot of upside, especially with Ford on board."

Arriving at Ford on January 11, 1965, Walsh says King was a welcome addition to the radio department. "The 8 Track was put into production too fast, in my opinion, and King was very instrumental in making sure the program rolled out as best as possible," he says. Initially, King was assigned a desk in the radio-engineering department, located on the third floor of Building 5. It was an open environment, dominated by rows and rows of wooden desks. Other activities in the building included chassis engineering and several large drafting rooms. With his wife and six children still in Waukegan, King leased an efficiency apartment on the second floor of a home in Dearborn. In early April, the family purchased a five-bedroom home in the village of Beverly Hills, a quiet suburb located some 20 miles northwest of downtown Detroit. "I actually never saw the house before we purchased it," Barbara says, "but I saw one of a similar style in West Bloomfield

Township when I visited in January, and it was very nice. Ford paid for the moving expenses, so we rented a large truck and loaded up the cars and moved. We had to do all the normal things, like find a private school for the children."

Jumping into the 8 Track program his first day, King represented the radio-engineering department at a launch meeting inside a conference room at the Ford Division offices at Rotunda and the Southfield freeway. In attendance was John J. Nevin Jr., a Ford product-planning manager who held an MBA from Harvard University, Motorola's Kusisto, and RCA's Tarr. Nevin, a chain-smoking, hard-driving executive who joined Ford in 1954, left the automaker in 1971 and went on to run Zenith Radio Corp., which sold televisions and other consumer electronics, and Firestone Tire & Rubber Co. After Firestone was sold, he held a controlling interest in Budget Rent A Car Corp. In a 1987 interview with *Businessweek* (now *Bloomberg Businessweek*), Nevin wasn't shy when asked to comment about the periodic disagreements that occur at any corporation, including several run-ins with Iacocca (who joined Ford in 1946 and was appointed vice president of the car and truck group in 1960). "The report that we had a lot of squabbles was absolutely correct," Nevin said of Iacocca.

After Tarr, the polished New York marketing executive, relayed that RCA was fully supportive of the program, the group reviewed a list of the challenges in adapting the 8 Track system to the automotive environment. The player and tapes had to operate over a wide temperature range, from -20 degrees to +140 degrees, including up to 95 percent humidity levels. The system also had to withstand vibrations and possible interference generated from other vehicle electrical systems. The player needed to be compact, both for installation within, or below, the dashboard. The list of components packed in the player was staggering: A manually tuned AM radio, a complete tape mechanism, stereo preamplifiers, two power amplifiers, and stereo volume, tone, and balance controls. A small door, which automatically covered the tape opening when the cartridge was removed, prevented dust, dirt, and cigarette ashes from getting into the player mechanism.

Another factor was the selection and location of the stereo speakers. More of an art than a science, given the technology of the day, at least two speakers are required for a stereo effect. Through various tests, it was determined that two speakers were optimal for small cars, while four speakers were installed in larger vehicles. As for placement, Ford determined speakers mounted on the doors, cowls, and under the instrument panel provided the best stereo effect. Of the three locations, door-mounted speakers provided the widest frequency range, since the door acted as an infinite baffle and extended the bass range. The cowl, or rear, speakers provided the second-best frequency range, while speakers mounted under the instrument panel proved to be adequate, but were not preferred.

Still, door-mounted speakers presented several challenges: The units had to be packaged to fit between the window and door operating mechanisms, and each speaker had to be moisture-resistant. A watertight seal was set between the speaker gasket and the door's inner panel to prevent any leakage from rain, melting snow, or car washes. To meet the time constraints, King and his team perfected an accelerated test to check the moisture-resistance of speaker samples from different suppliers, including rapid acceptance or rejection of production lots. In turn, a number of advancements to prevent moisture from reaching the speakers brought on the use of special gaskets, adhesives, and a partial plastic cone (the requirement of full bass response ruled out the use of full plastic or metallic covers). A drip shield was added, as well.

Taking into consideration colder climates, it was essential the speakers be moisture-free to prevent any lingering damage to sound quality. To ensure reliability, King and his team conducted a series of tests. First, an operating speaker was submerged in water for 30 minutes. After the unit was removed from the water bath and disconnected from the power source, it was immediately placed in a cold chamber at -20 degrees for a half-hour, after which power was reapplied and the speaker was expected to perform as if nothing had happened. The process was repeated in a hot chamber set at 130 degrees. As for the player, Ford and Motorola met a desired lifespan in excess of 1,500 hours under a range of environmental conditions. The cartridge, meanwhile, was set up to play for more than 500 hours, which was greater than the lifespan of an LP record.

Over the course of the winter and spring, King made sure the entire system overcame a combination of variables including temperature, humidity, voltage, mechanical shock, and on/off switching. Through it all, King maintained a work schedule of 60 hours per week, on average. "I was traveling between Motorola's Franklin Park headquarters, the tape player plant in Quincy, and RCA's record pressing plant in Indianapolis," King says. "For the times I went to Illinois, I would fly into Chicago O'Hare on a commercial flight. If I were going to Quincy, I would transfer to a Douglas DC-3 (built between 1936 and 1950). I made those trips more times than I care to remember; including over several weekends, where we would iron out all of the challenges. At the same time, I was writing the specs. It was hectic, to say the least."

## FITS AND STARTS

Soon after Lear Jet Stereo took possession of Motorola's office and plant facility on Lyndon, east of Schaefer, on Detroit's west side, Frank Schmidt, at Lear's direction, removed all of the clocks from the building. Lear, as was his custom, wanted the clocks taken down because he believed the staff would work longer hours, which is why nearly everyone who worked for Lear wore a wristwatch. "It was like a gambling casino … he didn't want you to know what time it was," Schmidt told *8TrackHeaven.com*.

As equipment, parts, and material arrived at the plant in early 1965, Schmidt worked on refining the 8 Track tape player and cartridge. While the Lear team claimed the first cartridges could withstand considerable pressure, the reality was quite different. "The initial Lear cartridges utilized a low-impact polystyrene plastic that was brittle," King recalls. "If you dropped it on the ground, it would often crack in several places. At my insistence, they went with medium-impact polystyrene plastic. They used plastic pins to hold the cartridge together, but they weren't strong enough, so they had to tighten those up. Later, they went with a screw, and they broke the head off to keep people from tampering with the cartridge."

In February 1965, while visiting RCA's record pressing plant in Indianapolis, where a corner of the facility was dedicated to the 8 Track program, Robert Moyer, manager of recording development engineering, commented to King that the early Lear cartridges were fine. "They're like a good bottle of Scotch," Moyer said. "Just don't drop them on the floor."

As for the player, Schmidt reconfigured the playback head so it read the tracks from top to bottom as Ford, Motorola, and RCA had requested. The inside-out motor stayed. "The Lear player had a fantastic motor; it might as well have been put on a lawn mower," says Barry Fone, owner of Barry's 8 Track Repair and The 8 Track Repair Center, both in Prescott Valley, Ariz. Fone, who earned an electronics degree from the U.S. Navy and has repaired scores of 8 Track tape players, says beyond the Lear motor, the early player left quite a bit to be desired, given the technical limitations of the day. "The motor ran a huge flywheel, which could easily be pulled out because gravity was about the only thing holding it in place," he says. "The motor and flywheel alone weighed five pounds, which was quite a bit of weight compared to other setups. Given they were working in 1964, when there wasn't a lot of technology, some things just didn't work right. When you opened up the case, there were often exposed wires that might short-circuit if the player was bounced around. Unfortunately, there wasn't a good track-changing mechanism. It actually utilized some fairly flimsy pieces of metal. Nowadays, I tell people to use the early Lear Jet Stereo 8 player as a decoration, because it's too expensive to repair."

Another early challenge was that the Lear team insisted on using a polyurethane pressure roller for inside the cartridge. The chief reason was cost. Up until that time, cured rubber rollers were commonly used in reel-to-reel players, but it was a more expensive offering. Schmidt told *8TrackHeaven.com* he "worked through about 30 or 40 different rubber and elastic compounds looking for the perfect material (for the pressure roller) that wouldn't flat-spot (leave a permanent indentation) in cold temperatures and wouldn't adhere to the tape in hot weather conditions. The one thing I wasn't looking for was a 30-year lifespan on the cartridges. You have to look at the materials that were available at that time. Very limited. Those rollers would barely last five years."

He was right. Before playing their 8 Track systems, many audiophiles today replace the pressure rollers inside Lear cartridges produced prior to 1970. "Later, RCA developed a hard plastic roller that most every cartridge manufacturer used," King says. "The Lear team might not have known their roller was too soft. When the tape sticks to the roller, it slows down the music to uncomfortable levels, or it can cause the tape to jam."

Next, the Lear team had to account for the graphite coating used on the backside of the magnetic tape. "On an endless-loop 8 Track, you're pulling tape off the center of the platform hub, and you're adding it to the outside of the platform hub, and you've got a speed difference," Schmidt told *8TrackHeaven.com*. "That roll of tape has to be loose because it's constantly sliding against itself. You lose a lot of that lubricant over a period of time. That's the black dust you get inside the tapes. Another thing that happened was with the binder that binds the graphite to the back of the Mylar tape. It starts to give out with age. There's nothing you can do about it. If you play that tape over and over, that tape is going to wear out … it's going to give up its lubricant and bind up tight. That's one of the limitations of the 8 Track system."

To keep graphite from building up on the tape head, Motorola developed a small plastic scraper next to the pressure roller that would remove any debris. In turn, RCA offered a Stereo 8 Head Cleaning Tape Cartridge, which proved to be the best-selling cartridge tape in its catalog by 1967, according to a *Billboard* interview with Ed Welker, RCA's manager of recording tape merchandising. The special cartridge contained a mildly abrasive tape that cleaned the playback head in 30 seconds or less. If tape-coating built up on the head, it could reduce high-frequency levels as well as volume. The cleaning cartridge, to be used after every 100 hours of player use, retailed for $1.95. RCA also sold a Stereo 8 Cartridge Tape Caddy — which could hold up to eight tapes — for $2.95.

The next challenge was to locate a machine that could re-record the reel-to-reel master tapes from RCA, which were one-inch across. The Lear master record machine used half-inch tape (which was reproduced on quarter-inch tape). "So where do you find this kind of professional equipment in Detroit? You have to remember that this was the '60s, and there wasn't a lot of equipment like this around," Schmidt recalled in *8TrackHeaven.com*. "Dick (Kraus, a Lear manager) finally found a place (north of downtown) ... only problem was that we had to do it late at night, usually after midnight. We would rent some time on their studio equipment because they weren't using it then. It was an outfit called Motown Records. ... (and) upstairs on the third floor ... they had the equipment, and they were kind enough to rent it to us late at night. I remember going down there lots of times and sitting around waiting while they were finishing up a session. There were some (ladies) singing ... they were called The Supremes. We used to sit around and talk with Berry Gordy (founder of Motown) all the time." As a result of the early relationship with Motown Records, Gordy was one of the first producers outside of RCA to license its music catalog to the 8 Track system.

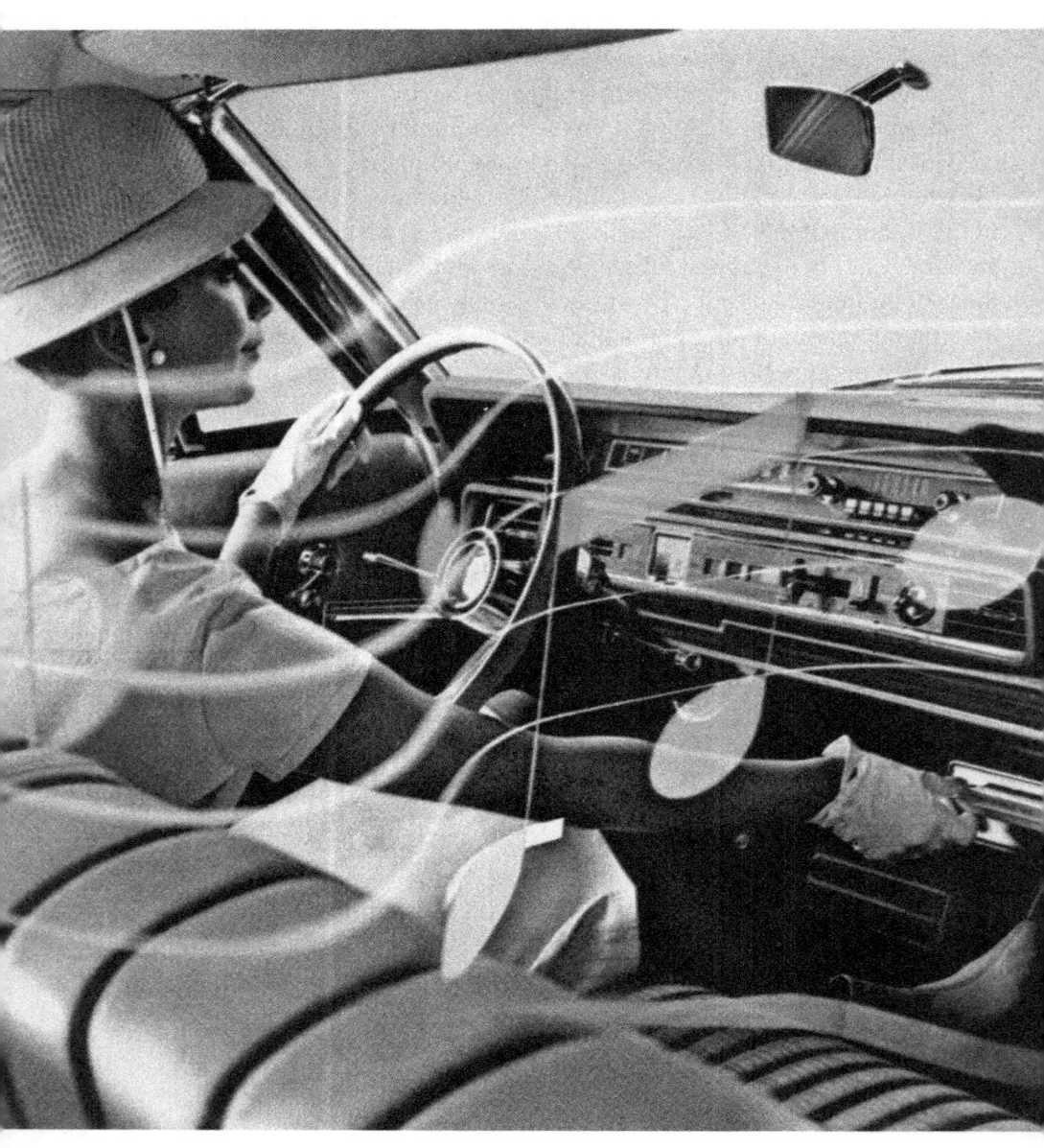

COURTESY OF THE BENSON FORD RESEARCH CENTER AT THE HENRY FORD

# ROLLOUT

## WHAT TO STANDARDIZE

THE 8 TRACK TAPE PROGRAM AT FORD was still under lock and key in the early spring of 1965. While there were rumblings in various trade publications that Bill Lear was working on an 8 Track system, few outside the recording industry knew the player and cartridge had left Lear's Wichita engineering lab. Even fewer knew Motorola had sold its Detroit operations to Lear for the purpose of jump-starting the production of parts and cartridges. "We were working in relative obscurity," King says. "It was easier that way. The more people you bring into a project like that, the more demands there are."

That all changed on April 3, 1965, when *Billboard*, in a front-page article, broke the news that Ford would offer Motorola 8 Track tape players in several of its 1966 vehicles, citing "highly placed sources in Detroit." The story came a week after the trade publication revealed RCA had reached a development deal with Lear. Under the terms of the agreement, RCA would make part of its record catalog available for car and home use in the Lear-developed 8 Track stereo player. The reporter, Lee Zhito, speculated the player might be compatible with 4 Track and 8 Track cartridges, not knowing Motorola's Kusisto and Ford had already nixed a dual player. As the story played out, RCA's competitors suddenly had to angle for position when it was revealed, via a source, that RCA would duplicate recordings for other labels (left unsaid, but understood, was a drive to license the technology and charge royalties).

Around that time, Alan Livingston, president of Capitol Records, released a letter to the media that sought to have the new industry standardized by a third party. He suggested the engineering committee of the Recording Industry Association of America (in which Capitol had great influence). The letter's original recipients were record and automotive leaders. He cited the industry's "costly and wasteful transition from 78 to 45 to 33 1/3 RPMs, a transition that even today is incomplete, with record players unnecessarily burdened with multi-speed mechanisms. Such a battle of the speeds in the newly emerging automobile stereo tape field can only penalize the public we all wish to serve." He added, "Whatever the playback system agreed upon, we believe it is essential that a tape cartridge which fits a Ford also fits a Chevrolet, Plymouth, and every other car make."

In a follow-up letter, Norman Racusin, division vice president and general manager at RCA, took issue with Livingston's position on the basis that the company, along with Lear, had already established its 8 Track cartridges as the industry standard (with assistance from Ford and Motorola). "The proposals enunciated by Mr. Livingston clearly suggest imposing on a dynamic industry a definition of what the elements of a stereo playback system should be," Racusin wrote. "In the near future, we expect to market pre-recorded tapes in continuous-loop cartridges for use in equipment which has been engineered in one form specifically for automobile installation and in other forms for home playback equipment. Many technical developments still lie ahead of us. It would be folly to freeze any elements of such developments and thus impede progress."

Walking a fine line, Lear said he supported Capitol Records' call for standardization; to a point. "The industry at the present stage does not have any tails to wag a big dog," he told *Billboard*. "I agree heartedly with Mr. Livingston. Now is the time to standardize. However, standardization must be based on economics and not history. The only standardization that we can hope for is in engineering standards, otherwise we can expect a stifling of the art."

Ten days later, on the morning of April 13, RCA invited record companies on the East Coast to a product exhibition in the Mercury Room at the New York Hilton. George Marek, vice president and general manager of RCA, kicked off the show by reminding everyone that 8 Track players and cartridges appealed to a young, vibrant audience (the median age of the country was 27). Nearly every American, not to mention a vast overseas market, was receptive to the concept, he said.

In the large room was Mike Stewart, executive vice president of United Artists

Records; William C. Gallagher, vice president of engineering and research development for Columbia Records; and George Lee, director of Eastern operations at Warner Bros. Records, among many others. Bob Schwartz, an executive at Laurie Records, told a *Billboard* reporter, "I was very impressed with the RCA Victor demonstration. Until now, I thought there were technological problems, but basically, I think they are solved." Lear was beaming. "I was extremely pleased by the manner in which RCA Victor conducted the demonstration, particularly by the fact that RCA didn't take a position of proprietorship or seek exclusivity," the inventor said. "Instead, RCA looked upon this as an extension of the entertainment art, and as such, is willing to cooperate in the fullest with the entire industry."

That open spirit failed to materialize during licensing discussions. Rather than pay royalties, many record companies took on production themselves, or tapped a fast-growing supplier base. If RCA or Lear sued, they'd take their chances in court, Campbell said. Among the early companies supplying magnetic tape, cartridges, or parts were Ampex and Orrtronics, both pioneers in the tape field. Ampex, through its subsidiary, United Stereo Tapes, announced it would make available its 1,300-album library for the Lear and Ford players. Orrtronics introduced its own player and cartridge, hoping to license its system to others. An underground cartridge market developed early on, with many labels subject to piracy.

Ampex, named after founder A.M. Poniatoff (his initials were combined with the first two letters of excellence), was one of the first U.S. companies to experiment in magnetic sound recording following World War II. Based in San Carlos, Calif., the company took "delivery" of a German-made Magnetophon that had been spirited out of Europe in 1946 by John T. Mullin, a liaison officer assigned to the U.S. Army Signal Corps. The Germans were using magnetic tape machines — based on the first known magnetic recording in 1898 by Danish engineer Valdemar Poulsen on his so-called telegraphone machine — to record interrogations of Allied military personnel who had become prisoners of war. Using editing and cutting techniques, the words were rearranged so that the prisoners said something entirely different than what they had, in fact, said. In addition to broadcasting the rearranged messages over German State Radio in a bid to sow doubt among the Allied forces, the Nazis reproduced speeches and messages from Adolf Hitler to make it appear as if the German leader was traveling extensively throughout Europe with no resistance from the American and British forces.

Mullin, who, after the war, became a professional recorder development manager at 3M, demonstrated an upgraded version of the Magnetophon in 1947 at the NBC-ABC studio in Hollywood to entertainer Bing Crosby and his production team. Crosby, one of the most popular American entertainers prior to his passing in 1977, was intent on recording his performances for broadcast on radio. The trouble was that recording limitations on standard lacquer discs produced a muddled sound. "Bing himself at times underwent considerable torture when the final disk assembly was played on the air coast-to-coast on the full ABC radio network," Mullin wrote in 1972. When Mullin played back Crosby's studio performance recorded on the Magnetophon, an engineer motioned him to stop the player after a few minutes. Mullin recalls he "pressed the 'STOP' button. There were surely no more than two seconds of silence, which seemed more like an eternity to me, and then, a shower of compliments," he recalled in an industry retrospective that appeared in *Billboard*. "One small machine, one of a pair, side by side on a makeshift table — the only two of their kind in the United States arranged to record and reproduce magnetic tape with such remarkable fidelity, that in a listening demonstration that lasted almost five minutes had upset the entire future of sound recording in this country."

Orrtronics had a similar history. In 1945, then Maj. J. Herbert Orr (he retired as a colonel) was a radio-engineering officer assigned to Gen. Eisenhower's staff at Supreme Headquarters Allied Expeditionary Forces. Legend has it Eisenhower wanted to address Europe through a mass radio broadcast as the war was waning. Using leftover German tape that had seemingly been erased, the future president quickly ended the practice because Hitler's muddled voice could be heard during the taped address. Gen. Eisenhower issued an edict that no more captured tape was to be used; while Orr was given the assignment of further developing magnetic tape with the assistance of captured German scientists.

Following the war, Orr returned to his native Alabama and quickly set up a magnetic recording tape company. Following several ventures, some of which were acquired by Ampex — which found it cheaper to buy Orr's plant rather than pay a markup — the inventor partnered with Cousino in Toledo to develop a line of 8 Track tape players and cartridges in 1966. The special cartridges were among the first that could be re-recorded.

Muntz, meanwhile, moved to grab as much market share as possible before the 8 Track industry got its footing. In April 1965, he dropped Stereo-Pak's price to $69.95. When he started, a less-technical system was available at $129.95, plus $25 for installation and four color-matched speakers. He also concluded a distribution deal with retail giant Montgomery Ward and signed an exclusive contract with Warner Bros.–Reprise. With the latter move, Muntz offered a total of 2,430 albums. "If there is significant market interest reflected in the Lear cartridge, we will issue 1,000 of our best-sellers in that form," Muntz told *Billboard*. He added the company was equipped to duplicate its titles on the 8 Track system, but there was no mention of duplicating players.

While other reports put the total market at 300,000 cartridges, Muntz said he had sold 2 million 4 Track cartridges in three years. By the end of 1965, he forecast sales of 8 million cartridges. The following month, he announced the introduction of a combination record player and 4 Track tape recorder which would retail for less than $300. He said RCA, Capitol, and Columbia, which had declined to license its music catalogs to him, "were losing $200,000 a year to bootleggers." By not extending their respective catalogs, the record companies "forced" him to develop the combined player and recorder, he added. According to the company, the three most copied artists by bootleggers were The Beatles, Elvis Presley, and Andy Williams.

At every turn, Muntz and his chief engineer, Eash, blasted the emerging 8 Track tape player, saying it would confuse consumers. "It is my feeling that the tape industry has been caught up in its own glamour," Eash wrote Livingston, of Capitol Records, in a letter that was made public. "Too many changes have been foisted off on the public that sound great in the laboratory but cannot be maintained in production."

At the same time, Muntz' main supplier, TelePro Industries Inc., filed suit against Lear in Wichita, claiming patent infringement. In response, Lear stated that his company automatically indemnified all users of his cartridges. What's more, for the first time, he revealed that Lear Jet Stereo had moved to Detroit and was on its way to delivering an initial order from RCA for 1 million cartridges by October.

To boost awareness, Lear and his sales team took the 8 Track program on the road. One of the first presentations was at the Capitol Records Tower in Hollywood. The West Coast challenger was the first to compete directly with the New York stalwarts: RCA, Columbia, and Decca, among others. Capitol enjoyed a rapid grassroots launch. Eager to start his own record company, songwriter Johnny Mercer formed an investment team in 1942 and opened a studio in Los Angeles. Drawing friends from the entertainment business, Capitol sold more than 40 million records in 1946. Artists included Nat King Cole, Tex Williams, Les Paul, Peggy Lee, and the Pied Pipers.

A few weeks after reviewing Lear's 8 Track player and cartridge, Livingston told *Billboard* he was "still examining" the system. Capitol, he said, also was reviewing the Orrtronics player. He added Muntz' Stereo-Pak was out of the running.

## SLEEPING GIANTS

The quest for the world's first playback tape system, dubbed the "automatic radio station," had its start in 1956. Around the time Chrysler was introducing the Highway Hi-Fi record players inside vehicles, Motorola showed off a 4 Track, continuous-loop tape system to Ford's executive team. "At the time we submitted our cartridge players to Ford, the product apparently was ahead of the market and there was interest generated, but because of the great economic risk, neither Ford nor the software people (record companies) would consider supporting the program," Kusisto reflected in the *Journal of the Audio Engineering Society* (October-November 1977). "They concluded that the market simply was not ready for a cartridge-type automotive player."

Left unsaid: Ford was spending millions of dollars on the Edsel, a new line of cars designed to compete against Oldsmobile, Buick, and Chrysler. Work began in 1955 on the program, code-named "E Car." To propel sales — four models were made available for the 1958 model year — Ford added nearly 1,200 dealers, giving it a total of 10,000 sales outlets (Ford, Mercury, Lincoln, Continental). By comparison, Chrysler had 10,000 dealers, while GM had 16,000 dealers.

Named after Edsel B. Ford, Henry Ford's only son, the Edsel line was designed to boost Ford's profits and market share. After driving Packard, Nash, and Hudson out of business or into the newly formed American Motors Corp. (Studebaker folded in 1966), Ford, buoyed by the enormous success of the Thunderbird in 1955, sought to grab market share from GM, Chrysler, and AMC.

The response from the buying public was less than overwhelming. Over the course of three model years, Ford sold north of 118,000 Edsels. Even before the line was discontinued in late 1959 (about 2,800 1960 models were built), the critics had a field day. As a case study, Ford conducted scant test marketing among consumers before the car was shipped to dealers. It also failed to set the Edsel apart from the Ford and Mercury lines in terms of styling, performance, and price, which tended to confuse consumers. Overall, the automaker lost $350 million on its $400 million investment. The disaster set the company back several years, and it only regained momentum with the success of the 1964 1/2 Ford Mustang.

With cash rolling into Ford's finance department and the overall economy improving in late 1964, the company's executive team became open to new investments. Soon after Lear introduced Ford II to the 8 Track tape player, Lear contacted Harry Tucker, an engineer in Ford's advanced concepts area who worked closely with Frey. The latter executive quickly green-lighted the program, even as outside engineering counsel recommended that "it was impossible to produce, within the time frame they were considering (nine months), 8 Track cartridges of recordings, maintain the fidelity and accuracy of head placement, and, of course, have the system live though the rugged environment of automotive application at a price that the consumer would be willing to pay," Kusisto wrote.

Despite the headwinds, Frey invited four competing electronics suppliers, including Motorola, to submit proposals for a tape-playing system with or without an integrated radio. Seeking to bridge the Muntz and Lear offerings, Motorola developed a compatible 4 Track and 8 Track player complemented by an AM radio. As the competition played out, Motorola was selected as the preferred developer, given the company's long association with the automaker and its proven ability to meet tight deadlines on so-called crash programs. The selection came as a relief to Kusisto and Motorola, as Ford had acquired electronics supplier Philco in 1961 in a bid to produce radio units internally.

It wasn't until 1964 that Philco started producing AM radios for Ford, along with Motorola and Bendix. As for Philco's future as a sole supplier of radios, the plan never fully played out, and Motorola continued to produce Ford units for another decade. Still, Motorola's management team was concerned that the 8 Track tape playback system was a major gamble. In fact, Motorola's consumer products division (Quasar) didn't give the program much chance of success. As the supplier of Motorola-branded home audio equipment and components, Quasar had a big say in whether to support the 8 Track program. The division's wait-and-see position didn't stop Kusisto, who had the ear of upper management because of the fact that his automotive products division produced private label radios for the auto industry — not to mention that it was by far the largest-volume producer of car radios.

Reflecting on the number of musical chairs in the deal, Kusisto said much of the program with Ford was done without finalized purchase orders. He had Frey's handshake approval to undertake the program, "and it was all that Elmer Wavering, general manager of Motorola's automotive products division (later to become Motorola vice chairman), required," Kusisto wrote. "When the purchase orders were assigned, the contract was agreed to without any finalized price of the tape-radio combination unit. Ford and Motorola had built enough trust in one another's integrity that that detail was unnecessary. It may sound unbelievable in this day and age that business was conducted that way, but this is a true story."

During the crash program, Motorola's 8 Track team quickly abandoned Lear's setup with its integrated direct-drive motor. The problem was the arrangement was likely to create serious problems in the field, meaning tape speed could fluctuate and impact the playback process. Working with Japanese suppliers, Motorola developed a belt drive that rotated the capstan at an even pace with the assistance of a transistor speed regulator. About halfway through the project (five months), Motorola had a working prototype of a combined 4 Track and 8 Track player that was integrated with an AM radio. Given the need to accommodate Muntz' 4 Track design, where a built-in pinch roller flipped up to create the necessary pressure to start the audio playback process, the slot for the 8 Track cartridge was set above the radio dial scale. That meant drivers might have trouble reading the dial when a cartridge was inserted into the playback slot.

After factoring in the engineering challenges and the amount of royalties to be owed on Muntz' patents, Kusisto requested Ford drop the 4 Track from consideration three months before launch. "We reviewed with Ford management the fact that we were running a tight deadline and that if we removed the engineers' insurance of 4 Track compatibility, we could then be assured of having 8 Track on stream through full and thorough commitment," Kusisto wrote. "After some days of deliberation, (Ford) came back and advised us to concentrate on 8 Track only."

From the start, weekly engineering meetings were held between Ford, Motorola, RCA, and Lear Jet. Ideas were pooled together as the team tackled the design, engineering, and assembly of a reliable, compact playback system that was so easy to use that a 3-year-old child could insert a tape without a problem.

The mood of those who touched the project, from the labs to the engineering departments to the executive wings, was a fusion of risk and excitement. No one talked openly about whether his career was on the line, but it was likely in the back of everyone's mind. Motorola was the obvious choice to produce the players, given the company had every motivation to take on a two-and-a-half-year project and whittle it down to nine months. By cornering the market on the introduction of the 8 Track tape player, Motorola saw a way to stay relevant within Ford, which at the time controlled 19.6 percent of the U.S. auto industry. "We took an enormous risk ... with an over-commitment in engineering budgets to meet the timetable," Kusisto wrote.

At the time Motorola was working with Ford, Kusisto and his team called on the competing automakers. General Motors was interested in the system, but wanted to wait to see how the 8 Track system was accepted in the marketplace. If the automaker decided to go forward, it planned to build its own units via its Delco Radio division. Eager to get as many customers as possible, Motorola offered its counsel and technical experience to the automaker at no charge. As it turned out, General Motors had already started to test and review the Lear player, according to *Billboard*. Meanwhile, Motorola convinced Borg Warner to distribute aftermarket 8 Track players, which would be installed below the dashboard, to Ford dealers across the country. The other automakers — Chrysler, American Motors, and Volkswagen — were intrigued but had reservations. Following GM's path, they took a wait-and-see position. Ford, Lear, and Motorola also made numerous visits to competing record companies in a bid to increase the library of musical offerings, namely pop tunes, Broadway shows, and operas.

As the 8 Track team raced to solve all of the technical problems, meet the summer 1965 deadline for delivery to Ford, and expand the market, Kusisto decided to do an end run around Motorola's Quasar division. "We met jointly with RCA principals to review the music and worked with RCA to develop the home market," Kusisto wrote. To save planning and development time, Motorola's automotive products division would manufacture the first home units. "Our proposal was selected with some trepidation," Kusisto wrote. Still, he had all the bases covered. Motorola would supply the auto and home players, RCA the music, Ford the initial customer base, and Lear the cartridges.

In mid-July, the fledgling industry received a shot in the arm. The Lear Jet team secured a distributor. Larry Finley's International Tape Cartridge Corp., with offices at 1290 Avenue of the Americas in New York, agreed to purchase 1 million cartridges. At the time, Finley estimated 300,000 4 Track cartridge units were in the marketplace, and he predicted "untold thousands of 8 Track" cartridges would be bought. A former nightclub owner in New York, Finley moved to Los Angeles in the 1930s and set up a line of jewelry stores. He went on to produce and host TV and radio shows before buying up the tape cartridge rights for close to 60 labels, including Dot, Audio-Fidelity, Horizon, Seeco, Tico, and Vee Jay. Among the formats represented: country, latin, classical, folk, rhythm and blues, jazz, and rock 'n' roll. Artists included Elvis Presley, John Lee Hooker, The Beatles, Little Richard, the Andrews Sisters, Jimmy Dorsey, Count Basie, and Louis Armstrong.

Mindful of the mistakes resulting from Chrysler's short-lived Highway Hi-Fi system, in which there was a limited selection of the special, 7-inch records available in stores and dealerships, Lear saw mass distribution as the launch pad for the industry. In late August, Finley announced his company, in partnership with National Mercantile Corp., would seek to stock pre-recorded 8 Track cartridges in Ford dealerships. The practice, known as rack jobbing, involved the placement of cartridge display racks. The two firms supplied the racks and cartridges on a purely consignment basis, maintained the inventory, and oversaw bookkeeping duties. After 30 days, the distributors would be reimbursed for the merchandise, less a fixed commission. At the outset, the cartridges retailed from $2.98 for a single LP to $9.98 for a double album. Some 650 single albums and 200 double LPs were offered for sale.

Finley, a master salesman, began penning a column for *Billboard*, simply titled: "Tape Cartridge Tips." In his first offering in early September 1965, Finley explained why the company's factory in Fairfield, N.J., had an armed guard at the entrance, along with a sign that read: "Positively No Admittance."

"If you were running our factory, you, too, would carefully guard the manner in which we master our tapes, as well as duplicate them for 4 Track and 8 Track cartridges," he wrote. "... There is more to it than just mastering and duplicating. There is a special way in which we keep our plant dust-free, our air conditioning at a certain set temperature, and actually maintain our premises as if it were an Aerospace Lab. When we say we give our cartridges the 'white glove' treatment, we are not kidding. In fact, we purchased an additional three gross (36 dozen) of white gloves this past week!"

As the various partners prepared a massive advertising and marketing blitz to introduce the new medium, Muntz dropped his cartridge prices to $2.98, in line with the lowest price of the Lear offering. Having introduced production efficiencies at his Van Nuys plant, which led to lower labor costs, Muntz claimed he had dropped his prices by 60 percent since 1963. He also said his Stereo-Paks were sold in 375 retail outlets, giving him 65 percent of the total cartridge industry at the time. While he knew the 8 Track tape player was coming, he didn't know the extent to which the new system would be marketed across a broad spectrum of media outlets.

## BUBBLE OF SOUND

Connie Francis couldn't believe her latest album, "Connie Francis Sings All Time International Hits," was playing inside the limousine as it whisked her from the airport. In August 1965, Francis had arrived in Las Vegas to begin a weeks-long engagement at the Sahara Hotel. The limo was equipped with one of Muntz' 4 Track tape players. "Hearing it in the car was just like riding in a 'bubble of sound,'" Francis wrote Finley. "I really didn't know that the latest albums were available on ITCC tape cartridges, and want to compliment you on being on the ball in the way you are running your company. Would you be kind enough to airmail four copies of this cartridge to me at the Sahara, and send a couple to

George Scheck in New York so that I may have them when I return?"

The following month, almost everyone in America would be introduced to the 8 Track tape player. Both Ford and RCA had taken out full-page ads in a host of automotive and music trade journals as well as popular consumer publications like *Life, Time, Newsweek, Sports Illustrated, Esquire, Playboy, Hot Rod, Motor Trend, Rod & Customs, Sports Car Craft, The Wall Street Journal, The New York Times,* and numerous regional newspapers. In addition, radio and TV commercials were aired on a variety of channels, including an NFL game telecast. Suppliers like ITCC, Motorola, and Scotch Tape took out advertisements, as well.

Meanwhile, Ford's automotive dealers, many of whom had never seen a tape playback unit and were skeptical that the new medium would be a hit, saw their fears subside at new model unveilings held throughout the country. As Ford personnel showed off 1966 Mustang, Thunderbird, LTD, Galaxie, Mercury, and Lincoln Continental models equipped with 8 Track tape players, the mood was festive and ambitious. In one showing in Detroit, 136 Mustangs were ordered on the spot, according to *Billboard*. John Gall, sales manager of Lear Jet Stereo, said dealers were ordering the system "in impressive numbers," and many saw a ready market for units installed beneath the dash. To help generate buzz and sales, RCA created an advertising campaign where distributors could compete to win one of eight new Mustangs, each equipped with an 8 Track tape player.

Initially, RCA delivered 100,000 cartridges to Ford dealers. Customers who ordered the optional player received a cartridge filled with a variety of music. Lear Jet Stereo was producing 2,000 cartridges a day at its plant in Detroit, and by midsummer planned to deliver 4 million units. At the same time, the 4 Track market received a shot in the arm as TelePro, which manufactured the Fidelipac cartridge, acquired a one-third interest in Muntz' former company, Muntz Auto Stereo. The outfit planned to ramp up cartridge production significantly, and hoped to ride the advertising campaign by Ford and RCA. Victor Muscat, president of TelePro's parent company, Defiance Industries Inc., told *Billboard*, "the child is growing larger than the product, and therefore should be spun off to grow on its own." Muntz' Stereo-Pak units also reportedly were selling well.

With consumers able to purchase 8 Track tape players from Ford in early October, Lear was experiencing pushback from the record industry. Under Lear's royalty offering, a record company would pay 15 cents per cartridge for up to 15 million units, 10 cents for the next 15 million units, and five cents after 30 million cartridges had been sold. Livingston, president of Capitol Records, ripped into the arrangement. "Why should a major record manufacturer with its vast catalog, representing an investment of millions, have to pay Bill Lear a royalty for each cartridge used?" he asked a *Billboard* reporter. "We have the LP and the 45 RPM license free; why should we support a new method which costs us money? What makes it worse is that these payments must go on in perpetuity."

Lear, who enjoyed mixing it up, didn't take kindly to Livingston's attack. "How cheap can you be?" he charged in *Billboard*. "We've invested $2.5 million in developing this system. Aren't we entitled to something for our efforts? It's incredible that the one man who championed the cause of standardization of cartridges in automobiles is now the lone holdout when it comes to accepting our system, which is emerging as the automotive industry's standard."

In a follow-up letter, Livingston agreed with Lear. "Of course we want him to show a return on his investment, but how long does our industry have to keep paying for his patents? Why doesn't he give us a cutoff date? Say after a certain number of cartridges, or after a reasonable period of time, he will turn over the system to the recording industry on a royalty-free basis. How dare any record company loan its tremendous catalog to the development of something which may someday become a sledgehammer over our heads?"

Lear didn't bend. "We've opened a new market for this industry," he countered. "This is a market which had not existed before. If Livingston wants to cash in on this market, he should be willing to pay a few pennies for the opportunity. The number of records one can use is governed to a great extent by the amount of time he is near his playback equipment. When you are driving — for business or pleasure — your turntable is idle. An idle turntable spells unsold records. This is music for 'people on the go,' whether they're in boats, planes, or automobiles. The cartridge allows recordings to accompany them wherever they go."

As the battle raged on, Capitol, Columbia, and Decca took their time weighing the practicality of entering the market. But the more the companies waited, the more time RCA had to grab market share. Seizing upon the opportunity, RCA rolled out a major incentive program, including offering auto dealers a catalog of songs available on 8 Track that could be sent to consumers. The company also offered a leather carrying case, called a caddy, which could hold 10 cartridges. If a consumer bought six cartridges at once, the case was free. "The 8 Track tape player was a big hit, and in some cases, as strange as it may sound, the player sold the car," says Tarik Daoud, former fleet manager at Al Long Ford in Warren, Mich. (he acquired the dealership in 1972). "The customer wanted it, and nearly everyone at the dealership wanted one. I had players in my cars, and one in the house. The sound was great, and the recording artists had another medium to sell their songs. I don't think the people who paid for commercials were very happy, because people weren't listening to the radio as often."

But others had to be convinced of magnetic tape's viability. David Usher, a longtime producer who served as president of Argo Records in Chicago from 1958 to 1960, says it took him months to cajole his friend, Mickey Shorr, to carry 8 Track tape players and cartridges in his chain of audio electronic stores in metro Detroit.

"I knew Mickey since the early 1950s when he was a deejay at CKLW (800 AM) in Windsor, Ontario," says Usher, chairman of Marine Pollution Control in Detroit. "We became good friends, and I was producing for artists like Dizzy Gillespie. Well, I had a second home near one of Mickey's stores on Woodward Avenue in Royal Oak, (Michigan). We were such good friends that when he was on vacation I would pick up the money from the store each day and deposit it in the bank.

"In early 1966, I said, 'Mickey, you've got to get into magnetic tape.' But he said he was too busy. So I bought an 8 Track tape player and tried it under my dash, and it worked really well. So I gave it to Mickey and said, 'Try it.' A few days later, I asked him how he liked the player. Turns out he had sold the player right away, and he wanted more. So I got him some more, and it really launched the business to another level. Mickey had more work than he could handle, and it allowed him to open more stores. After a few months, he and I started getting the players from Lear Jet Stereo when they were based on Detroit's west side. He sold them like hotcakes."

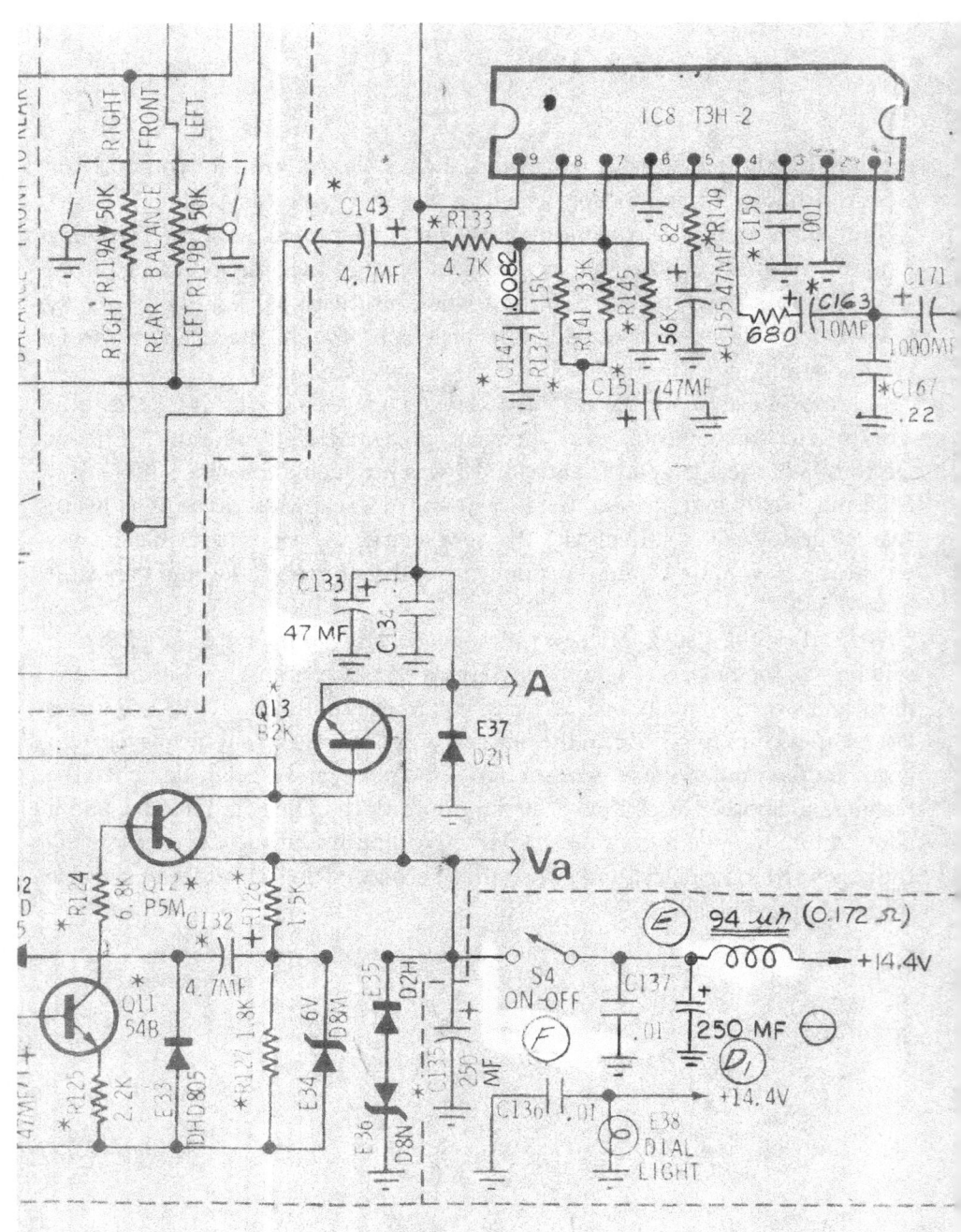

A circuit diagram for the Quadrosonic 8 Track tape player, from Nov. 21, 1974.

*Ford Motor Company,*

PRODUCT DEVELOPMENT GROUP

INCHES ON ORIGINAL FORMAT 'D'

| 1 | 2 | 3 | 4 | 5 |
|---|---|---|---|---|

DO NOT SCALE     ABOVE SCALE FOR REFERENCE ONLY

REF

| DRAWN BY | COMPLETED | CHECKED | APPROVED | SCALE |
|---|---|---|---|---|
| D Benart | 11-21-74 | Hartes | SA Alloy | NONE |

## UNLESS OTHERWISE SPECIFIED

DIMENSIONS ARE IN INCHES          3rd ANGLE PROJECTION
MACHINED DIM +                    STAMPED DIM ±
ANGULAR DIM ±                     DIM ±

MATERIAL

_____

APP J. Sin

DATE 11-21-74

NAME SKETCH-SCHEMATIC DIAGRAM RADIO
RECEIVER (AM/FM/MPX/QUADRASONIC TAPE)

NO. MS-D6SA-2713-AA

99

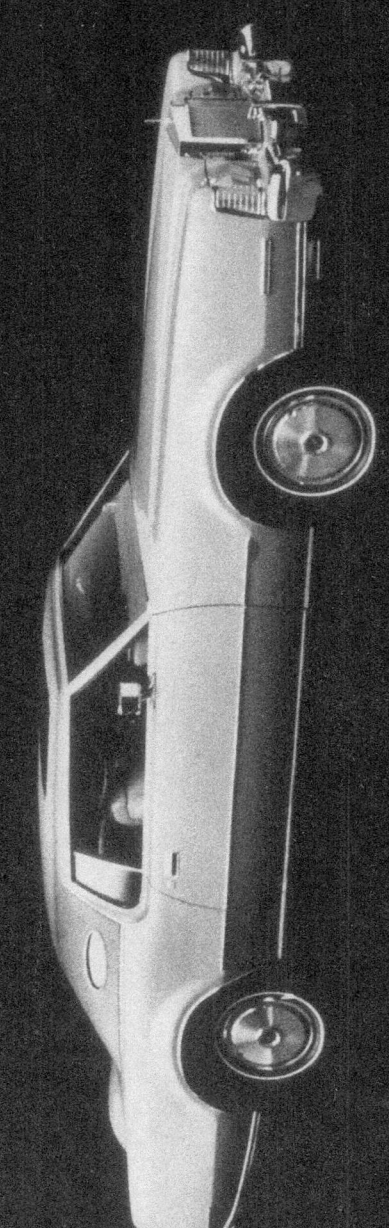

OCTOBER 3, 1975

1st QUAD 8 TAPE SYSTEM IN U.S. AUTOMOBILE

The 1976 Lincoln Mark IV was one of the first vehicles to offer a Quadrasonic 8 Track tape player.

QUADRASONIC TAPE
AM STEREO FM

The 1976 Lincoln Mark IV was one of the first vehicles to offer a Quadrasonic 8 Track tape player.

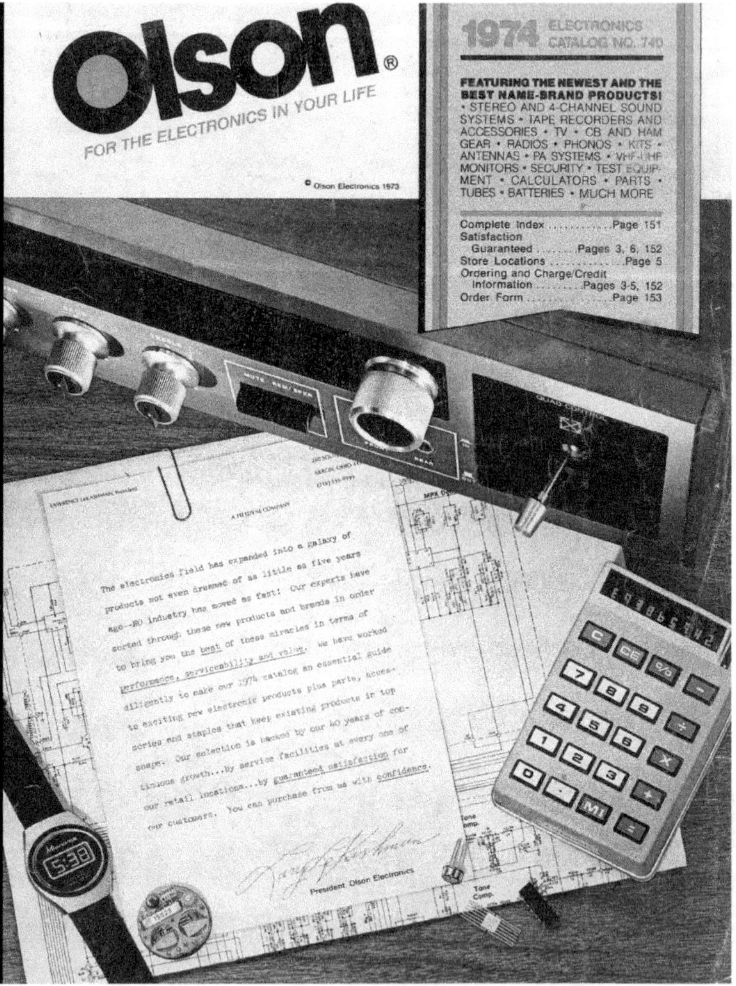

# CRAIG. 8-Track Auto Entertainment Systems

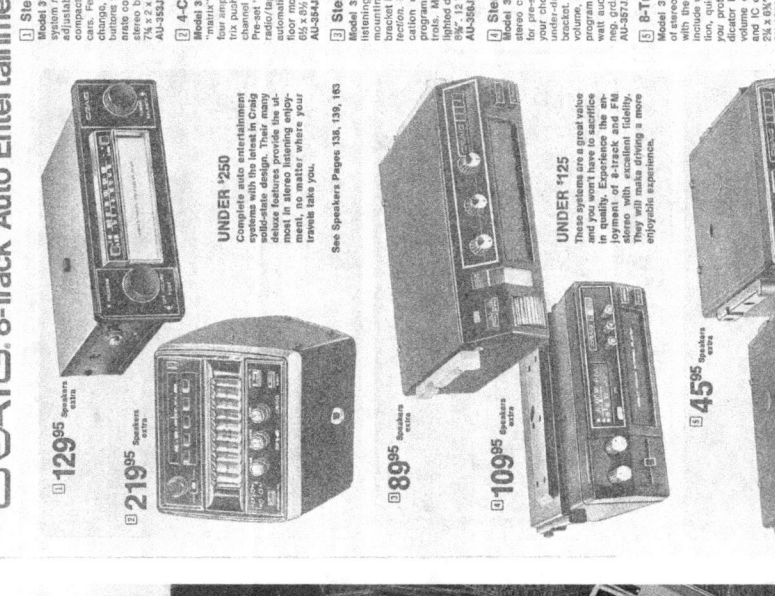

**[1] Stereo Player/AM-FM Radio**
Model 3128. Complete stereo entertainment system incorporates the best of your car. The adjustable controls and convenient are in a compact size make it adaptable to most cars. Features instant radio/tape program change, sensitive AM/FM radio with push-buttons for stereo/mono, FM and AM, separate controls for tuning, volume, tone, and balance/fader. A-C-4-watt audio output. 7⅝ x 2 x 9⅝". For 12 VDC neg. grd.
AU-3533. Shpg. wt. 4¾ lbs. ... 129.9

**[2] 4-Channel Player/FM Radio**
Model 3133. Designed for TRUE 4-channel "matrix" 4-channel, or stereo listening with four amplifiers. Discrete cartridges, matrix pushbutton for stereo or synthesized 4-channel from cartridges or FM broadcasts. Pre-set "5-button" FM stereo tuning, instant radio/tape mode change, "eject" button, automatic stereo switching, AFC. Include floor mount bracket. 24-watt audio output. 6½ x 8½ x 9¾". 12 VDC neg. grd.
AU-3544. Shpg. wt. 9¾ lbs. ... 219.9

**[3] Stereo Player/FM Radio**
Model 3139. Excellent value for your auto listening enjoyment. The under-the-dash mounting system has a quick release bracket for your convenience and theft protection. Features fast forward for easy location of tape selections. Multicolored program indicator lights and balance controls. Manual tuning dial for FM radio lighted dial. 14-watt audio output. 8⅝ x 2¾ neg. grd.
AU-3563. Shpg. wt. 6 lbs. ... 89.9

**[4] Stereo Player/FM Radio**
Model 3137. Many features! Plays 8-track stereo cartridges—includes 2-pushbutton for pre-set AM or more FM stations with your choice, PLUS manual tuning. Secure under-dash mounting with quick release bracket. Also includes sliding-type tone, volume, and balance controls, multicolored program indicator lights. Fast forward. 14-watt audio output. 7⅝ x 2⅞ x 9¾". 12 VDC neg. grd.
AU-3573. Shpg. wt. 9 lbs. ... 109.9

**[5] 8-Track Stereo Player**
Model 3144. Enjoy the listening pleasure of stereo in the privacy of your own car and with the convenience of 8-track. Features under-dash mounting for maximum installation, quick release bracket for take it with you protection. Also includes program indicator lights, tone switch, separate slide volume controls, manual channel selection and IC circuitry. 14-watt audio output. 8½ x 2⅝ x 9¼". 12 VDC neg. grd.
AU-3584. Shpg. wt. 6 lbs. ... 65.9

**[6] 8-Track "Deluxe" Player**
Model 3155. You get the auto entertainment of 8-track with many of the features found in more expensive models. The fast forward control lets you locate your favorite cartridge recordings easily and quickly. Secure under-dash mounting with quick release bracket. Also includes sliding-type volume and tone controls, multicolored program lights. 14-watt audio output. 4⅝ x 2⅝ x 9⅝". 12 VDC neg. grd.
AU-3564. Shpg. wt. 9 lbs. ... 54

---

**UNDER $250**
Complete auto entertainment systems with the latest in Craig solid-state design. Their many deluxe features provide the utmost in stereo listening enjoyment, no matter where your travels take you.

See Speakers Pages 136, 139, 163

**UNDER $125**
These systems are a great value in quality. You won't have to sacrifice in quality. Experience the enjoyment of 4-track and FM stereo with excellent fidelity. They will make driving a more enjoyable experience.

**UNDER $65**
These are Craig's lowest priced 8-track players. They're great values for the budget-minded individual who still wants excellent sound and cartridge convenience in his own car.

[1] $129.95 *Speakers extra*
[2] $219.95 *Speakers extra*
[3] $89.95 *Speakers extra*
[4] $109.95 *Speakers extra*
[5] $45.95 *Speakers extra*
[6] $54.95 *Speakers extra*

144 Olson

---

## olson®
FOR THE ELECTRONICS IN YOUR LIFE

© Olson Electronics 1973

**1974** ELECTRONICS CATALOG NO. 740

**FEATURING THE NEWEST AND THE BEST NAME-BRAND PRODUCTS!**
• STEREO AND 4-CHANNEL SOUND SYSTEMS • TAPE RECORDERS AND ACCESSORIES • TV • CB AND HAM GEAR • RADIOS • PHONOS • KITS • ANTENNAS • PA SYSTEMS • VHF-UHF MONITORS • SECURITY/TEST EQUIP • METERS • MODULES • PARTS • TUBES • BATTERIES • MUCH MORE

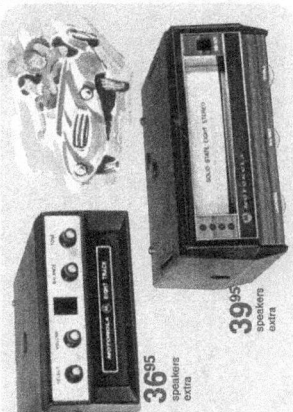
A selection of 8 Track tape players and accessories in Olson, a catalog for electronics enthusiasts, from 1974.

Various 8 Track tape players and cartridge carousels from the King Family Collection (Photos courtesy of Nick Martines).

Various 8 Track tape players and cartridge carousels from the King Family Collection (Photos courtesy of Nick Martines).

Various 8 Track tape players and cartridge carousels from the King Family Collection (Photos courtesy of Nick Martines).

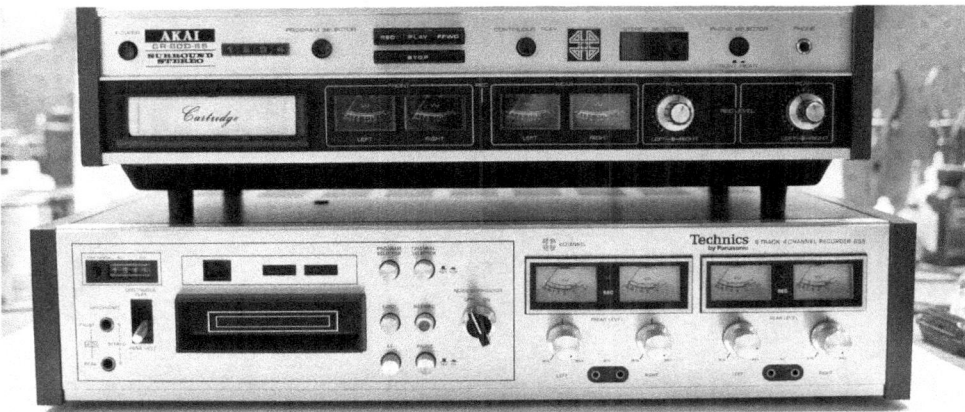

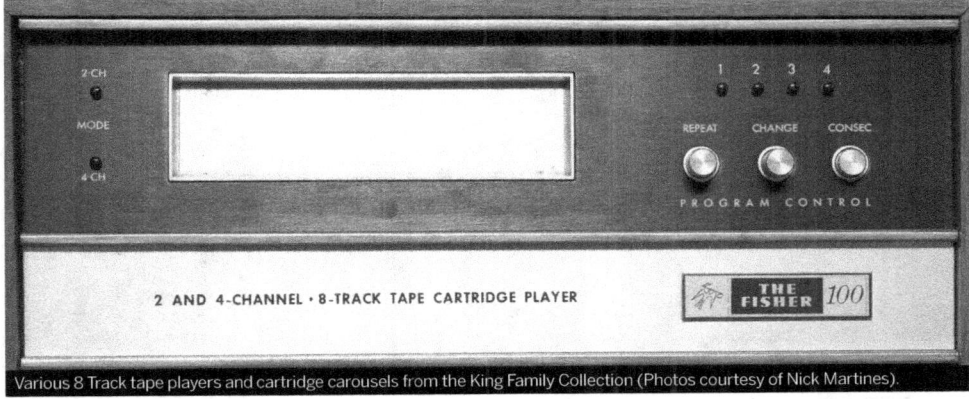

Various 8 Track tape players and cartridge carousels from the King Family Collection (Photos courtesy of Nick Martines).

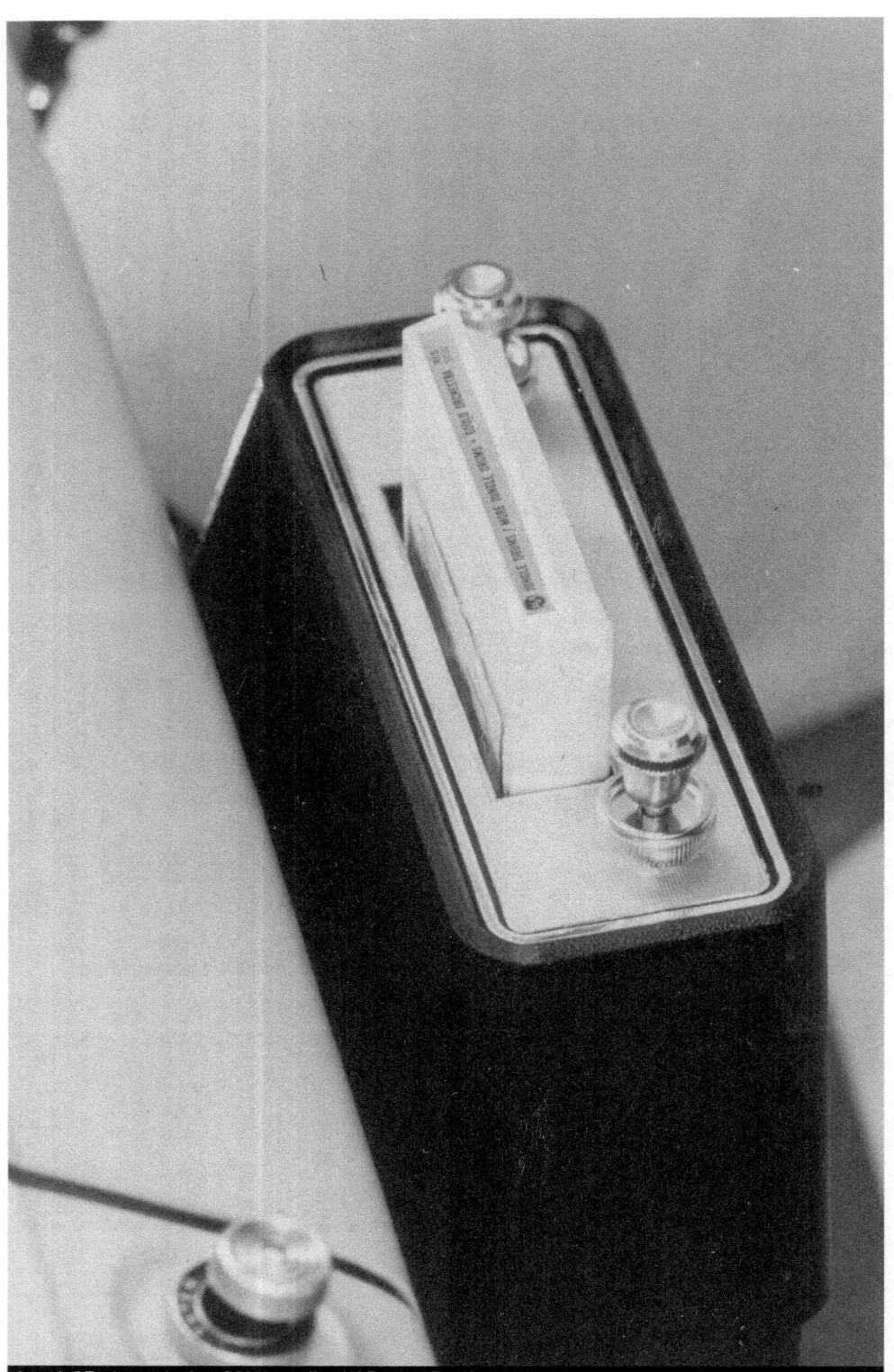

An early 8 Track cartridge from RCA Victor offered "A Family of Fine Music" selections.

Upon King's retirement from Ford in 1997, his colleagues presented him with a drawing of what life would be like following a 32-year career at the automaker (Photo courtesy of Nick Martines).

# INTERNATIONAL TAPE ASSOCIATION, INC.

## Presented To

# FORD MOTOR COMPANY

In recognition of the 10th Anniversary of their introduction of 8-Track Stereo in automobiles. Their imagination and confidence led the way for others to follow.

Ford's success in creating consumer awareness of this innovative entertainment medium led to the development of the home market. Because of the Ford Motor Company, our industry has prospered and profited by the sale of tape decks, magnetic tape, and cartridge parts.

Ford, a major contributor to making tape a household word throughout the world.

NEW YORK CITY                    OCTOBER 8, 1975

In 1976, to mark the 10-year anniversary of the 8 Track tape player, King, on behalf of Ford Motor Co., was presented with a commemorative plaque (Photo courtesy of Nick Martines).

John Jackson, chairman of the International Tape Association's 1976 conference at the Marriott's Essex House in New York City, and Larry Finley, ITA chairman, present King with the plaque (as pictured in the ITA newsletter).

*Billboard* Oct. 25, 1975

## Tape/Audio/Vide

VCT Press photos

**ITA HIGHLIGHTS**—At recent New York International Tape Assn. "semi-technical" seminar, big hit was quad mix-down demo by Joe Lopes of RCA Records, left, and Jack Richardson of Toronto's Nimbus 9 Productions. At right, ITA executive director Larry Finley reads plaque presented to John King of Ford Motor Co., center, "for creating consumer acceptance of the auto 8-track player" a decade ago, as Duane Windahl of 3M Co. looks on.

From *Billboard* magazine (October 25, 1975).

Jim and Earl Muntz sold thousands of 4 Track players, mostly on the West Coast.

Earl Muntz was a prolific entrepreneur, having started more than a dozen businesses.

COURTESY OF MUNTZ FAMILY COLLECTION

**DRIVE ROD**
A capstone in the player
propelled the tape
and the roller through
the cartridge.

# 07

# AWAY
# WE GO

## COMMUTATOR AND TRANSISTOR

THE OFFICIAL LAUNCH OF THE 8 TRACK tape player was not without a hitch, though no one knew it at the time. Certain 1966 models rolling down the line in late August of 1965 at Ford's Wixom Assembly Plant (Lincoln, Thunderbird), as well as at other plants (Mustang, large Ford, and large Mercury vehicles), saw the installation of the Motorola AM Radio and Stereo Tape Player. In the first year, according to Ford personnel, the 125,000 players installed within, or underneath, dashboards made for a 25 percent installation rate. Touted in print advertisements, there was no mistaking "The Motorola Car Entertainment Center" that would let you "play all your favorite tapes." Of all of the ads ordered up, perhaps the best of the bunch was several dozen three-quarter-page ads in *Billboard*, where Ford, RCA, Motorola, and Lear Stereo reached a cross-section of automotive and trade executives as well as young adults and teens.

Finley, the largest aftermarket distributor (and music licensor) of cartridges, revealed in mid-October of 1965 that due to the massive advertising program, the "industry has advanced as much in these past 10 days as it has in the past three years. We are being swamped with orders from all over the country, not only for the Lear Stereo 8 cartridge, but for the 4 Track as well as the Orrtronics

8 Track cartridges." He added, "The advertising is helping all phases of the automobile stereo field. Our factory is working three shifts a day, seven days a week, and all orders are being filled on a 'first come, first served' basis."

Max Fine, who presided over Gem Electronics, a 30-store chain based in New York that had outlets spread across the country, told *Billboard* that the pre-recorded tape industry had taken four years to find a footing. Initially confined to classical and operatic offerings, sales began to climb significantly following the introduction of rock 'n' roll and pop tapes in 1965. The company's offering of pre-recorded tapes was poised for a breakout year, he projected. "When fathers come by the store to pick up some electronic equipment, parts, or tapes for themselves, they buy these rock 'n' roll tapes for their children," he said. "The tapes have a different image in the father's mind than do records."

Apart from advances in electronics, the pre-recorded tape industry, both for reel-to-reel and cartridge offerings, found a growing market of younger buyers. According to *Billboard's* annual survey of record retailers (1964), buyers age 30 and younger purchased 87.7 percent of all singles and 68.5 percent of all LPs. Digging deeper into the data, 67.8 percent of singles were bought by buyers 19 years old and younger. It was safe to assume the tape cartridge industry would ride the coattails of these young buyers as they moved into the job market, when the purchase of an 8 Track tape player and multiple cartridges would be in easier reach.

At the same time, buyers of new Ford, Mercury, and Lincoln cars were driving sales of cartridges in major markets, since the respective dealers located in small towns across America had yet to take delivery of the full catalog of music due to supply challenges. Throughout the remainder of 1965 and into the following year, record store owners in New York, Chicago, Los Angeles, and other major markets heard the same story from visitors: Customers complaining, sometimes bitterly, of their inability to purchase cartridges in their hometowns. As a result, customers typically would take several record store catalogs home in order to pass them out to family and friends. As a result, mail order sales proved to be a healthy part of a record store's business in the early months of the 8 Track rollout.

Seeing the writing on the wall, Muntz announced in late October of 1965 that he had orchestrated a deal to make his catalog available on Lear cartridges. Because Muntz' 4 Track cartridges could not be played in the Motorola players, he planned to invest $50,000 on new equipment. The new cartridges would be ready for sale in the first quarter. At the time of the announcement, Muntz claimed he controlled 70 percent of the cartridge business and 60 percent of the player market.

A look at the numbers backed up Muntz' claim, but many of his cartridge sales came from smaller labels. "The amount of royalty we pay establishes the price of the cartridge," he said in a *Billboard* front-page story. "Some companies are better off to take a lower royalty rate and get the (sales) volume."

He said his biggest seller was Johnny Rivers, who the following year would record "Secret Agent Man," one of several hit songs by the artist. Muntz also relied on a number of older recordings from the Platters, Frank Sinatra, Sammy Davis Jr., and Dinah Washington. What he lacked was access to new releases from RCA, Columbia, United Artists, and Warner Brothers.

Two weeks later, Jim Gall, marketing director of the Lear Jet Stereo Division, claimed Muntz was out of bounds. Even though Muntz said he had 100 sample Lear cartridges that he was testing, Gall argued the aftermarket dealer had not signed a purchase order. *Billboard* reported that Lear had no intention of bringing Muntz and his 4 Track system into the fold. With a deal in hand with Finley, and the Lear team negotiating with Columbia and Decca — even Capitol would be using the system "despite what Alan Livingston says" — Gall said 8 Track was well on its way to being the preferred tape playback system. In response, Muntz was quick to discount Lear's advances in the marketplace. In an unattributed article, a *Billboard* writer described Muntz' position: "With typical Muntz aplomb, he cited Lear as being 'desperate for music.'"

A week later, Ford's Frey revealed that one in five Thunderbirds were being equipped with an optional 8 Track tape player. If the installation rate held through the 1966 model year, close to 14,000 players would have been ordered for the Thunderbird, making it more popular than air-conditioning. "If there is that much enthusiasm for a so-called 'luxury' option like an air-conditioner, we are very excited about the outlook for the much less expensive Stereosonic tape player unit," Frey said. He added RCA was making plans to double its tape library for the following year.

Technically, sound produced from the 4 Track tape player, in which the tape traveled at 3 3/4 inches per second, was slightly better than an 8 Track player. But the difference in sound was difficult for most people to discern, given the mix of noises produced by a vehicle, along with surrounding traffic.

Beyond the fanfare and tape cartridge distribution challenges, Motorola's engineering team soon picked up on the looming arrival of a glitch in their 8 Track home and auto cartridge players, made worse by the sound it produced. Because the early motors were designed to operate at 12 volts, the power was delivered by way of a commutator (which connected to the rotating portion of the motor).

While the engineering team believed a transistor circuitry would help extend the life of the contacts on the motor, they unknowingly triggered a reaction. As more data arrived from Motorola's life-test cycles, it revealed that carbon from the motor brushes was prone to pack up in the grooves of the commutator after a few hundred hours of playing time.

The buildup of fine carbon material, in some cases, would trigger a signal that would cause the transistor to engage fully. As a result, the motor would operate at two to three times the normal speed, producing what was best described as "chipmunk sounds." To fix the problem, the engineering team changed several resistors in the motor circuitry to prevent the transistor from receiving a signal to run at top speed. Following that, Motorola set about routing dozens of affected customers to its service stations for the necessary repairs.

Another fix, which was adopted early in the launch of the cartridge, was the addition of a scraper. To keep the magnetic tape from sticking to a rubberized or hard plastic pressure roller, especially in warm weather (heat cycles, in engineer parlance), a small plastic scraper was added inside each new cartridge.

Motorola and Lear also moved to fix another early problem, dubbed "cross talk" — or an occurrence when a slight movement of the tape relative to the position of the tape-reading head could cause two tracks to be played at once, or reduce the sound level. In addition, in early models, a cartridge could move slightly after being inserted into a player — another contributing factor to cross talk. To prevent the cartridge from potentially controlling the movement of the tape, Motorola added a plastic insert to the tape guide to ensure that the tape "broke" over the head correctly and didn't roll up or down. For its part, Lear opened up the guides in the cartridge so that the player would control the tape path. Down the road, in some 1967 and later models, Motorola changed the support of the tape head from a radial motion (where the head changed its angle to the tape) to one that used a parallelogram mechanism, meaning the head was parallel to the tape.

## PARITY, PLEASE

By mid-November 1965, industry leaders were eager to standardize. Peter Fabri, president of Musictapes in Chicago, noted in *Billboard* that the industry was not prepared to support as many as eight separate pre-recorded tape systems, including two from Europe. He told *Billboard* that standardization was a prerequisite to mass commercial appeal, rather than the alternative where "we confuse the customer and butcher the market." He went on to predict the 4 Track tape player would gain dominance, but he left an opening for 8 Track. "We are flexible enough to go either way," he said. "Eight or 4 Track could usher us into a pre-recorded tape Utopia; if only we get together."

By Thanksgiving, 8 Track would get a tremendous boost. Chrysler and Lear Stereo announced a partnership in which the automaker, as soon as the first quarter, would offer aftermarket 8 Track tape players at its 10,000 dealerships. It was a departure from Ford's integration of a player into a vehicle dashboard, but it was the quickest way for Chrysler to enter the market. Had Chrysler followed Ford's lead and installed players on vehicles rolling down the assembly line, it would have taken until late 1966 (1967 model year) to ramp up the program. Chrysler, through its Mopar brand, would make the Lear-built 8 Track unit available in all of its vehicles — Chrysler, Plymouth, Dodge, and Imperial — integrated with an AM radio or as a standalone unit.

At the same time, Sears, the largest retailer in America, announced it had signed a deal to offer Lear's combination AM radio and 8 Track tape player. While the terms of the deal were expensive —Lear charged a royalty of 15 cents per cartridge, while Orrtronics, a competitor, was available at a royalty rate of five cents per cartridge — Sears was eager to establish its network of more than 4,000 stores as the top supplier of pre-recorded tape cartridges.

Next up, B.F. Goodrich, an auto aftermarket parts supplier, said it would test-market the sale of Lear and Muntz cartridges among its 450 company-owned stores. The company also looked to expand the sale of cartridges to some 5,000 gas stations that offered retail goods. Chuck Schultz, a Goodrich buyer merchandiser, said the move was spurred by Chrysler's decision to, like Ford, offer the 8 Track tape player. Finley's ITCC, which now represented 42 record labels, said it would make available all of its titles to 8 Track. The company was experiencing revenue growth of 30 percent per month. With the entry of Chrysler into the market, Finley told *Billboard* the industry had arrived. "When a Texas account offers you a guarantee of 40,000 cartridges a month, it's here," he penned. At the time, Texas was the second-largest cartridge market, behind California. Finley said he planned to offer 150 new releases a month by the following spring.

David Handleman, whose family operated Detroit-based Handleman Co. — at one time the largest independent wholesale merchandiser of records in North America — says myriad forces that neatly culminated in a few short years propelled the cartridge industry. "The world was clearly ready for a mobile music experience by the mid-1960s," Handleman recalls. "In the late 1950s you had Elvis Presley, who arguably was the first superstar entertainer, followed by the emergence of rock 'n' roll and The Beatles. But the one record that revolutionized people's thinking of what a popular album could do in terms of sales was a comedy record called 'The First Family.' No one had seen anything like it."

Recorded on the evening of October 22, 1962, in front of a live studio audience, the album was a good-natured parody of the Kennedy family, with actors offering impersonations of the president and his extensive network of family and friends. Issued by Cadence Records, the album sold 7.5 million copies in some six weeks, making it the fastest-selling record in the history of the entertainment industry at that time. Kennedy, portrayed by comedian and impersonator Vaughn Meader, was amused by the album and gave out copies as holiday gifts to family members and friends. At a Democratic Convention speech in early 1963, Kennedy quipped: "Vaughn Meader was busy tonight, so I came myself."

Handleman says the success of the album help lay the groundwork for the cartridge industry. At the time, record players were becoming more mobile with the introduction of smaller units that could be carried easily between a family room and a bedroom or basement.

"We were approached by Irwin Tarr (from RCA) to distribute pre-recorded music (records) in supermarkets in the 1950s, which we agreed to do," Handleman says. The company, founded in the 1930s, initially sold health and beauty aids and moved into the music business two decades later. "At the time, records were sold in five and dime stores like Woolworth's and Kresge's," he says. "Then the department stores came, like Montgomery Ward, Sears, and Kmart, which debuted in 1962 under the direction of Harry Cunningham, who really was the architect of the modern department store. We went on to distribute 8 Track cartridges, and it was my father who thought to distribute them in Firestone stores in addition to the normal channels."

As the industry soon learned, the duplication of inventory became a major issue, Handleman says. "In addition to records, you had to provide for multiple cartridges, and that impacted floor space," he says. "That meant only the big acts were sold (initially), as there wasn't enough room to carry a deep inventory, especially in smaller stores."

As 8 Track found consumer acceptance, the Muntz Stereo-Pak player was losing ground on various fronts. In late 1965, Al Bennett, president of Hollywood-based Liberty Records, announced he was dropping the 4 Track in favor of the 8 Track, saying the latter system "is the ultimate answer." Bennett also suggested the company might enter the cartridge business by investing $75,000 in mastering and duplicating investment. A few months later, in March 1966,

Liberty announced it had signed a deal with Lear to provide Liberty music on 8 Track cartridges, dubbed "Trak Pak." Bennett also announced he would produce titles for 4 Track players but, like the 8 Track, the company would handle its own distribution.

Liberty Records got its start in 1955, but three years later found itself near bankruptcy, although it did manage to eke out Ross Bagdasarian Sr.'s "The Chipmunk Song (Christmas Don't Be Late)." The actor, songwriter, singer, and record producer handled all of the voices on the song, including the father, David Seville, as well as the three chipmunks, Alvin, Simon, and Theodore (named for three Liberty executives). Numerous other Chipmunk material followed, including songs, movies, and TV shows. Ross Bagdasarian Jr. said his father was down to his last $200 in 1958 when he bought a special tape recorder for $190 that operated at varying speeds, which allowed him to produce the Chipmunk voices. Perhaps Bagdasarian Sr.'s biggest claim to fame was a minor role in Alfred Hitchcock's 1954 thriller "Rear Window," in which he portrayed a down-on-his-luck singing piano player who writes and performs the song, "Lisa," during the film (early in the film, Hitchcock makes his "cameo appearance" inside the singer's apartment). The label represented dozens of other artists, including, for a time, Henry Mancini, Julie London, and Eddie Cochran.

In early December 1965, Kapp Records in New York made headlines after it signed a distribution deal with ITCC. The pact brought the number of albums in ITCC's cartridge catalog to more than 1,000 LPs. Finley, of ITCC, said the most-requested artist in the cartridge industry was Kapp recording artist Jack Jones, who had a number of hit singles in the 1950s and 1960s, and went on to record the theme song to the popular TV show "The Love Boat," which aired for nine seasons on ABC.

A week later, Goodyear Tire and Rubber Co. and Borg-Warner announced they would each enter the cartridge industry; Goodyear planned to offer the Orrtronics 8 Track tape player and cartridges (which weren't compatible with the Lear and Motorola players), while Borg-Warner would offer 4 Track players and cartridges at its automotive aftermarket stores. Motown Records also announced it would make available five of its artists for 8 Track, in a deal with RCA. The artists were Marvin Gaye, the Miracles, the Supremes, the Temptations, and the Four Tops. In addition, Ampex and Motorola were beginning work on cartridge players for the home.

Anticipating greater things for the industry, at the dawn of 1966 a Tennessee furniture manufacturer, Berkline Corp., rolled out the "Stereolounger," a leather recliner equipped with a Lear player concealed in one of the armrests. Speakers were set on either side of the recliner, below the seat. There also was a jack system that would allow users to plug in a TV, a radio, or a Hi-Fi system. Available in seven styles, the Stereolounger, which included the player and one sample cartridge from ITCC, retailed for $239.

Walking through the 39th annual National Auto Accessories Exposition at Chicago's McCormick Place in late January 1966, Finley said he was inundated with cartridge requests. "Walking down to the Tenna booth, we watched Don Slack demonstrate their new 8 Track Lear configuration auto unit," Finley wrote. "Here I politely removed our competitor's cartridge and inserted ITCC's Tijuana Brass on the A&M label. Within minutes, the booth was so crowded that we hardly had shoulder room."

As Finley worked the convention floor, he left cartridges at every stop, finishing up at the Borg-Warner exhibit. "Vincent Vecchione was hosting one of the largest crowds at the show," Finley recalled. "The Borg-Warner mono unit was being played in front while, behind the beautiful stage setting, there was a cutaway of an automobile; the dashboard mounted with the new Borg-Warner Fidelipac configuration and the 8 Track Lear configuration. Vince also helped himself to what was left of our cartridges, so, not having any cartridges left, I decided to leave the show."

In a report following the event, *Billboard's* Paul Zakaras observed that many conventioneers "seemed to be somewhat in the dark" about the numerous players, and whether 4 Track and 8 Track cartridges were compatible. "The dozen new tape units demonstrated by eight firms received such an enthusiastic response that it now seems inevitable that the auto-electronics industry will plunge into the music business in a big way," he reported. On a related point, Zakaras wrote that some distributors had yet to make up their minds on selling players and cartridges at record stores. "Installation of our units is so simple that the average car owner will be able to do it himself in a matter of minutes," claimed Vecchione, manager of Borg-Warner's consumer electronics division.

## MR. DEEDS FROM DAYTON

As Ford, Chrysler, Muntz, and Orrtronics waited to see if Capitol, Columbia, and Decca would make their respective titles available to the cartridge industry, General Motors took the good part of a year to develop and test several players, both for a 4 Track and an 8 Track. Lear, Muntz, Motorola, and Orrtronics tried to convince the automaker to install their respective players, but GM, then the largest company in the world, wasn't all that interested. The reason can best be attributed to Col. Edward Andrews Deeds, an electrical engineer, inventor, and industrialist from Dayton, Ohio.

While at National Cash Register Co., one of the nation's first modern enterprises, he undertook in 1899 the design and development of the company's first electric plant. In addition, he wired the factory floor. Following a promotion to director of development and construction in 1903, Deeds hired a promising electrical engineer, Charles F. Kettering, to develop a cash register powered by electricity. In 1906, Deeds and Kettering debuted the new register, which set the company up for decades of growth and prosperity. Two years later, having difficulty constructing a car from a kit, Deeds asked Kettering and a handful of key personnel for assistance. Rather than rely on the kit's ignition system, Kettering, working in Deeds' barn, developed a high-energy ignition unit by tapping his knowledge from the motorized cash register. In 1909, Deeds and Kettering established Dayton Engineering Laboratories Co., or Delco for short.

From there, Henry Leland, who headed up Cadillac Motor Car Co., ordered 5,000 Delco ignition sets and had them readied for introduction on 1910 models. A year later, Kettering introduced an electric starter, first installed on Cadillacs for the 1912 model year. Soon after, the electric starter became an industry standard, and Deeds and Kettering became millionaires.

By 1912, Kettering was working full time at Delco, where Deeds joined him in 1915. A year later, the company was acquired by United Motors Corp., which was formed by General Motors' founder William C. Durant. United, with its multiple companies, developed and built parts for various automakers until General Motors acquired the entire enterprise in 1918 for $45 million. It was the start of a remarkable career for Kettering, who went on to head GM's research division from 1920 to 1947. He, along with others at GM, sought to have a hand in nearly every aspect of automobile manufacturing including materials, parts, production, and distribution. The vertically integrated operation proved to be highly efficient for its time, given concern over sourcing raw materials and components at affordable prices and in sufficient volumes. GM, under the direction of Alfred B. Sloan (at various times GM's president, CEO, and chairman from 1923 to 1956), did more to embrace vertical integration than any other global corporation. The in-house business model endured for decades.

By 1966, it was a foregone conclusion that GM was working on a tape player for its automobile lines, but the media had a hard time discerning the level of investment, the scope of research, and when, if at all, an 8 Track unit would be introduced. One reason for the lack of information was that research was being conducted at Delco's radio division in Kokomo, Ind. (now Delphi), where few reporters roamed. As news trickled out in early 1966, GM "sources" confirmed that Delco would introduce an 8 Track tape player similar in design and operation to the Lear and Motorola units. GM's Chevrolet, Pontiac, and Oldsmobile divisions would be the first to offer an optional player on their 1967 models, with Buick and Cadillac to follow for the 1968 model year. At the same time, American Motors and Volkswagen entered the picture, with Motorola supplying players for both automakers, while Ford announced it would make the 8 Track available on all of its 1967 models.

In the development of the Delco player, William Caldwell, head of customer research, said a "trick of the trade" was a shakedown procedure in which each unit was run for several hours before being shipped to assembly plants and dealership outlets. "Multiband auto radio sales (AM/FM), FM stereo sales, and tape installations are all increasing rapidly," he said. "People are in their cars a great deal more and simply want to be entertained. The trend to suburban living, longer vacations, and traffic tie-ups in metropolitan areas probably all has a bearing on this. With 40 to 80 minutes of pure pleasure listening available on one cartridge, the average businessman can be home before he has to change tapes."

While Lear Jet Stereo was disappointed to have missed out on supplying players to GM, it wasn't shy about the overall prospects of the 8 Track relative to the 4 Track. In one recurring print ad that debuted in spring 1966, the company touted that its 8 Track system offered up to "an hour and 20 minutes of stereo play on each cartridge; twice as much as 4 Track cartridges of comparable size." What's more, the ad highlighted that in "the big showdown between 8 Track and 4 Track stereo, more than 40 of the nation's top recording companies as well as the three major car manufacturers have decided in favor of our 8 Track cartridge system. How do you like them apples?"

For his part, Muntz took out display ads in auto magazines and *The Los Angeles Times*, and distributed direct-mail fliers. By the end of the company's fiscal year (October 31, 1965), national sales manager Sy Fralick reported $3.4 million in gross earnings. He added Stereo-Pak was on pace to register a projected volume of more than $8 million in profits in 1966. Recognizing the financial challenge of reaching millions of consumers, Muntz admitted to *Billboard* that a great deal of the demand for his offerings was the result of Ford's massive advertising program.

By the end of the first quarter of 1966, Muntz said his manufacturing operation had expanded to 48,000 square feet, from 9,000 square feet in 1963. He added the company's products were sold in 800 retail outlets. In comparison, the Big Three automakers and their respective dealership networks represented more than 30,000 sales outlets spread across major cities and small towns.

Competing against giants, Muntz proved there was consumer demand for pre-recorded tapes, but even he couldn't have predicted how popular the medium would become. On the front page of *Billboard* on March 5, 1966, Columbia, Capital, and Decca announced they would license their respective catalogs to the 8 Track system (Decca also would support 4 Track). Clive J. Davis, Columbia's administrative vice president, said the company "felt a keen obligation not to be swayed by the hoopla and publicity whipped up by those who may have other interests at stake," reported Mike Gross. "What we have been concerned about as a record company is the necessary technical refinement of the art, the market potential of this business, its profitability, and whether it is (a) substantial business." He added the record label had been studying the various cartridge offerings for almost a year.

At Capitol, Livingston — who, in the spring of 1965, had called for standardization — said the label was now convinced that the 8 Track would be around for the foreseeable future. Noting the entry of Ford, GM, and Chrysler into the market, he said the medium was coming together. "A compatible system is one of the keys to success for the tape cartridge," Livingston told Gross. "Without it there would only have been chaos, much the same as when the record industry created the battle of the record speeds." His colleague, Stanley M. Gortikov, president of Capitol Records Distributing Corp., in the same article listed several challenges awaiting the cartridge industry — a higher price relative to records, new inventory to be introduced and distributed, limited musical selections, and the impact on radio listening habits.

As Columbia, Decca, and Capitol ramped up their respective operations, Monument Records, a label that got its start in 1958 and was named after the Washington Monument, said it would establish its own tape operation by the end of 1966. The label's artists included Roy Orbison, Willie Nelson, Dolly Parton, Kris Kristofferson, and The Velvets.

At RCA, Sarnoff reported the company's profit of $82.5 million in 1964, based on $1.8 billion in revenue, would be further boosted by its pioneering efforts in the 8 Track cartridge segment. It was the 15th consecutive quarter in which the corporation's profits were higher than the comparable period a year earlier. The company, he said, had already invested $1.5 million to help bring the 8 Track cartridge to market, including a new manufacturing and production center, packaging, and duplication equipment. He added record stores were just now getting on board with the new product.

Already, several merchants were providing prospective customers with test drives, where they could experience the 8 Track on the open road. "We will loan them a Lincoln to drive around and listen to the system," said Don Shore, a record-store owner in Los Angeles. "They can ride all day if they wish. They leave their car with us, of course." In typical Texas fashion, Dynamic Devices in San Antonio claimed to offer the largest stereo tape library in the world. The company's 26,000-square-foot store included an area that could undertake as many as 12 installations at a time. In addition to displaying 12,000 different cartridges, the store doubled the number of soundproof listening booths to six in the spring of 1966; it said the booths were in continuous use.

In Boston, car radio pioneer Automatic Radio Corp. operated three shifts a day, and more than 400 workers were dedicated to the cartridge industry. "If we work for the next 10 years we may finally catch up with our present orders and the potential to come," said David Nager, sales manager of the company's consumer products division, in *Billboard*. "With 70 million cars on the road the potential is unlimited. This accessory has the most potential of any accessory since the automobile itself. People are going crazy to buy this unit. It has caught the fancy of the motorist like nothing else ever has, and it will be many years until we have even begun to supply the demand."

On the supply side, Amerline Corp. in Chicago, supported by Motorola, developed a cartridge which it claimed to have "100 percent tape lock" that would do away with tape spillage, loose turns, and tape hang-up problems. The units also included "screw lock construction," which would ensure alignment of the tape path relative to the playback head, as well as a precision pressure roller that would deliver "low-flutter performance." The company was already a longtime supplier of computer tape reels to IBM, Honeywell, Burroughs, and Ampex.

In Los Angeles, Magnetic Tape Duplicators, founded in 1953, announced in the spring of 1966 that it had invested $100,000 to duplicate pre-recorded music on 8 Track for several record labels, including Angel, Reprise, Capitol, Columbia, and Dot. One innovation in the company's production area was a conveyor belt that carried duplicated tapes to a loading area on the second floor. Overall, the company projected it would acquire 75,000 miles of recording tape in 1966. In a sign of the times, a company official, in a *Billboard* interview, declined to name the engineering team for fear competitors would lure them away.

Stretching the medium beyond entertainment, Musictapes in Chicago, which in late 1965 had predicted the 4 Track would beat out the 8 Track, changed its tune. The company, which reproduced and marketed tape cartridges for 400 accounts, moved to branch out into other applications. "Commuting businessmen can listen to sales training manuals and other taped materials while driving to and from the office," said Peter Fabri, Musictapes' president. "Tapes will instruct salesmen in their cars between calls. Car tapes will carry industry news, trade association reports, and legal and medical journal excerpts for busy professionals as they drive." To bring the point home, Fabri told *Billboard* that actor Eddie Albert used tape cartridges to brush up on his lines for the TV show, "Green Acres."

Riding the wave of options, Lear Jet was set to introduce, by the 1966 holiday season, a portable cartridge tape player powered by electricity or rechargeable batteries. With two speakers, it was the first unit to allow people to listen to music while walking, sitting on a beach, or relaxing on a patio lounger. At the same time, two Lear Jet executives left the firm and started Rael Inc. (Lear spelled backward), a 12,000-square-foot 8 Track duplicating firm in Troy, a suburb of Detroit.

"We are working with a new type of high-speed duplicating equipment and we have been accepting quite a few orders," said Rael President Dick Krause. "However, we will make no exaggerated claims. Before the end of March we will have duplicated only about 2,000 8 Track tapes. In April, we will produce 4,000 to 6,000 (tapes) and we will begin turning out 60,000 (cartridges) monthly in May. Eventually, after we are able to become automated, we should be capable of (producing) about 200,000 (tapes) per month." Krause added the recent announcements about large catalogs of music dedicated to the 8 Track "do not mean anything yet because lack of equipment and proper (duplicating) facilities has made it impossible to produce any significant number of tapes."

By summer, the company had switched its name to Stereodyne. In addition to manufacturing 4 Track and 8 Track cartridges, the firm produced point-of-sale materials and shipping containers. In September 1966, the company became the exclusive cartridge supplier to Motown Records.

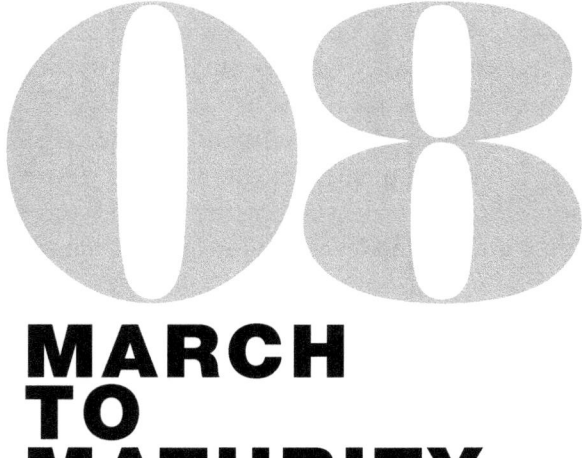

# MARCH TO MATURITY

## STAY TUNED

AS THE 8 TRACK TAPE INDUSTRY RACED to balance limited supply with seemingly unlimited demand, there was no shortage of distribution ideas. Finley, of ITCC, told *Billboard* he was developing a vending machine that would hold 120 tape cartridges. The telephone booth-sized machines, each equipped with a bill scanner, would be placed in gas stations, department stores, and other high-traffic retail locations. The report preceded by a week *Billboard's* new department feature, called Tape Cartridge News, which debuted on March 12, 1966. In an ad touting the upcoming weekly section, the trade publication said the department would provide "vital news coverage and product highlights on the multimillion-dollar industry with the multibillion-dollar future."

In one of the first reports, Livingston, of Capitol Records, had since softened his stance on standardization. Saying there were more orders "than we can manufacture at the outset," he said he was pleased that the Big Three automakers were setting the industry pace. "By the fall of 1967 we will have an indication of the real acceptance of the auto buyer," he said. "It will take one year to know the future for cartridges. We know there will be lots of business because there are lots of pipelines to fill."

The inaugural tape cartridge report also carried a small item about Phyllis Morris, a furniture designer and manufacturer based in Beverly Hills, Calif. The company was developing a line of bedroom night tables, end tables, and coffee tables equipped with 8 Track tape players and speakers (hidden from view).

On the international front, Mexico, England, Germany, Switzerland, Holland, and France started to offer 8 Track cartridge playback systems. Just as with most installations in the U.S., players were mounted beneath dashboards, often below the central column area (accommodating either left-hand or right-hand steering wheels). In a bid to stir up interest in the new medium among record stores, Finley decried the practice of tape cartridges selling for high prices in popular tourist spots. "What kills me is that Americans go to Acapulco for a vacation and pay $14.50 for a cartridge which can be bought for $5.95, only it's not available in their (hometown)," he said. "This shows that the record dealers aren't getting as involved with cartridges as they should."

In May 1966, with the growing popularity of the 4 Track and 8 Track, more American radio stations began to offer a request format. Popular in smaller markets, where a listener would call in and request his or her favorite song, the format moved to urban markets in Los Angeles, San Diego, Buffalo, and Denver, among others. In Minneapolis, KDWB used what it called an "electronic secretary" to process listener requests. As calls came in, a pre-recorded voice would instruct a caller to leave their name, age, and hometown, along with the name of a song and the artist. As *Billboard* described the process inside the studio, "Girls, listening in over a loudspeaker, type up a list and take this list directly into the DJ who's on the air, allowing the station to keep up-to-the-minute on records."

Charles Brown, the station's program director, said the format generated new listeners and encouraged longtime fans to be part of the action. "The word 're-quest' is a magic word," he said. "All of a sudden our DJs are enthusiastic again. They're totally involved. We're much more on top of hit records as a result of this all-request format. Another great thing is that we've found once people commit themselves with a request, they stay tuned. It reminds you of the old days of radio in smaller markets."

Dubbed You-Asked-For-It Radio, some stations had to place limits on the format. WYSL in Buffalo generated so many requests early on — some 26,000 calls a day — that the phone company sent word that the practice was overloading lines during peak hours of the day (10-11 a.m. and 2:30-4:30 p.m.). At KROY in Sacramento, the station was fielding more than 50,000 calls a day, while in Denver, the local phone company told KIMN that 70,000 daily calls could not get through to the switchboard, even with the station's three lines. And at KFRC in San Francisco, station manager Tom Rounds said the phone company told the station to drop the format because of high call volumes.

The emerging cartridge industry generated countless headlines around the country. More than 250 newspapers picked up a May 1966 article by AP entertainment writer Bob Thomas, titled, "It Looks Like Car Stereo Has Proved Its Point." Other headlines at the time highlighted the excitement of the new medium: "Musically We Drive Along," "Autos Have Instant Music," "Dashboard Stereo Talk of the Industry," and "Auto Players Whittle Those Freeway Miles."

As the newspaper stories played out, the Electronic Industries Association gathered for two days in Washington, D.C., to develop a uniform exterior dimension for tape cartridges. Led by RCA chief technical administrator H.E. Roys, the special ad hoc industry committee drew officials from Ford, Chrysler, Lear Jet, Kodak, 3M, Delco, Muntz, and many others. The order of the day, as one executive put it, was "making the socket fit the bulb."

Part of the discussions centered on the pending formation of the U.S. Department of Transportation, approved by Congress on October 15, 1966, partly in response to activist lawyer Ralph Nader's book, *Unsafe at Any Speed*. In what would soon become a major agency of the transportation department, predecessors of the National Highway Traffic Safety Administration set numerous federal motor vehicle safety standards and regulations, including a rule that mandated the installation of seat belts in every vehicle. Other regulations addressed dashboard safety including padding, a reduction of hard surfaces, and no sharp corners. The 8 Track tape cartridge, as a result, was redesigned with rounded edges and corners. "The new design actually saved the cartridge producers money over time," King says. "With the old design, a manufacturer placed two stickers on a cartridge, one on the front and one on top. But with the rounded corners, a single, wrap-around sticker was developed."

At the same time, the recording industry began to issue standards to prevent publishers from losing control of their copyrights. And it was no wonder, as RCA announced on April 25, 1966, that it had "passed the million mark in (cartridge) production." Jim Gall, Lear Jet Stereo Division's director of marketing, predicted the industry would produce some 2 million 8 Track tape players by the end of 1967. Taking things further, the owners of those 2 million players would, on average, purchase 20 cartridges, or 40 million units overall, he said. In addition to traditional albums, educational series, and background music, Gall said movie soundtracks recorded on 8 Track tape were exhibiting tremendous sales. The most popular soundtracks included "The Sound of Music," "Dr. Zhivago," "The Singing Nun," "South Pacific," and "Bye Bye Birdie."

With Motorola enjoying strong sales, its fear of competition within Ford began to materialize. In June 1966, Philco announced it would begin building various 8 Track tape players for the home, including one with an integrated AM/FM radio. It was a foregone conclusion, at least among Motorola management, that Philco would begin building 8 Track units for vehicles. Founded as Helios Electric Co. in 1892, the company sold various lamps and what it called a "Socket Power Battery Eliminator," which allowed radio owners to operate their units via an electric wall socket. In the late 1920s, the company, then called Philadelphia Storage Battery Co., or Philco, entered the radio market. It soon became one of the top three radio producers in the U.S., and went on to manufacture air conditioners, refrigerators, washing machines, dryers, electric ranges, and televisions.

The company also worked with NASA to develop and introduce a global tracking station network for the Project Mercury space program, as well as the subsequent Gemini and Apollo missions. As Philco prepared for the launch of its 8 Track unit for the home, Ford, sensing Motorola's unease, decided to break up the competitive tension. Motorola would supply the home unit for Philco and, in return, Ford would market the players at its dealer locations.

The start of the summer season brought the presentation of the first "gold cartridge award," a takeoff of the highly popular gold record. While the criteria for a gold record was $1 million in sales (at wholesale value), Finley and the ITCC lowered the threshold to $250,000 in cartridge retail sales. At a ceremony inside the 5,100-seat Birmingham Auditorium in Birmingham, Ala., Finley presented the gold cartridge award to Herb Alpert and the Tijuana Brass. Following the award ceremony, the band played two sold-out shows, one at 7 p.m. and the other at 9:45 p.m. Finley, ever the showman, saw the award as "an innovation which, we believe, will become an accepted practice in the field."

In its advertisements that early summer, RCA touted the fact that it was offering more music in the 8 Track format than other record company. "Summer travel and vacation time is the peak selling season for 8 Track stereo cartridge tape; developed and introduced by RCA Victor," the ad said. "For the name of your nearest RCA Stereo 8 distributor, write RCA Stereo 8, Dept. EW, 155 E. 24th St., New York, NY 10010."

During the same period, Lear Jet announced it had signed deals with close to 60 distributors, giving it reach across 90 percent of the country. Eager to stay in front of consumers, Muntz made plans to debut a 12 Track player, which he said would meld the best of the 4 Track and 8 Track. The used-car dealer and entrepreneur also brought out a portable 4 Track player that would retail for $34.95. The device played an LP cartridge as well as a new product, a single tape similar to a single record. The first song to be offered: Frank Sinatra's "Strangers in the Night." Selling for 98 cents, the single cartridge would complement future releases to be based on Billboard's Hot 100 chart. Muntz added he was entertaining several offers for his company, including $6 million from Gulf & Western. The deal was never consummated. He also said Warner Bros. had made inquiries for Stereo-Pak, although the talks quickly ended with no sale.

In late June, *Billboard*, along with Survey Service, released a report that showed more than half of the nation's record dealers were carrying 4 Track and 8 Track cartridges. Concurrently, Finley said he was thankful for the report, having bet every nickel he had to start ITCC (in late 1964, after he retired from MGM Records as special director of sales). He saw the potential for tape cartridges and sought to secure licenses from the recording industry, the first of which was consummated with Tape Handling Co. in Fairfield, N.J. After meeting with some 100 "bankers, investment houses, and record companies," Finley convinced H. Earl Smalley, chairman of Dextra Corp. in Miami, to back ITCC. Housed in a 400-square-foot office in New York at the start, the company grew to 11 employees by the time larger quarters were needed. Initial projections for $35,000 in monthly revenue were quickly supplemented by reality. During the first month, ITCC booked more than $1 million in orders.

A natural salesman, Finley (born Lawrence Finkelstein in 1913) was raised in Syracuse, N.Y. He attended Crouse College, but didn't last long on campus. "I was bounced out after three months," Finley recalled in June 1966, on the first anniversary of founding ITCC. "I staged a one-man demonstration to protest the school's ban on cars on campus by driving a car up the steps of Crouse College. Some 50 youngsters were going to join me, but when none of them showed up, I did it myself." Following his dismissal, Finley became the leader of a local band and decided to shorten his name so it would fit on theater marquees.

Later, after starting up a jewelry store in Los Angeles, Finley formed the Progressive Broadcasting System in 1950 to complement his investments in radio and television, as well as three large ballrooms in L.A. "But my timing was bad," he recalled in a *Billboard* interview. "In January 1951, I woke up one morning to find I had lost $500,000 of my own money. I was wiped out.

"At this time, I was on the Friars Club board, sitting with all the top people of the entertainment business, but I was pumping gas in San Fernando Valley for

$60 a week from midnight to 6 a.m. No one knew of this in Hollywood until one morning Jack Broder, who owned the film production company Realart Pictures, drove in for gas. He was shocked. He had seen me at a Friars board meeting only a few hours before." Taking Broder's offer of a co-producer position, Finley slowly rebuilt his finances. In the early 1960s, after overcoming a serious illness, he worked for several record companies before ITCC was formed.

Never short of promotional ideas, Finley and his team developed "tapemobiles" from converted Greyhound buses. Inside the traveling showrooms, ITCC personnel would call on record store owners, electronics dealers, and other retail distributors around the country. The tapemobiles were loaded with the latest playback equipment from Lear Jet, Orrtronics, and Muntz.

As the industry played out, there seemed to be no shortage of display ideas. Motown Records unveiled a packaging concept in which a thin book was attached to the top of a cartridge. The flip-top setup provided space for song lyrics, contest information for fan clubs, and artist bios. When several such cartridges were stacked on end, the company noted the lineup would look like a row of neatly arranged pocket books. Despite a falloff in auto sales in mid-1966 and into 1967, factory-installed and aftermarket players enjoyed strong sales. Many dealers reported the 4 Track, with its lower price, was favored among younger buyers, while the 8 Track appealed to mature audiences that appreciated longer playing times, ease of use, and dashboard integration.

Prepping for GM's entry into the 8 Track field, as well as rising sales of units for the home, Finley said he would increase cartridge production by five percent per month starting in July, but the 4 Track would be kept at present levels. "There's no question that 8 Track will dominate the home market," Finley wrote. "This should happen about the first of the year (1967)." In the same *Billboard* article, Muntz, commenting on Admiral and Zenith's entry into the home 8 Track market, said 4 Track would continue to lead the pack. "You call them 'big boys,' I call them babes in the woods," he said of the TV giants. "I'll continue to outsell all of the combined with my 4 Track players (though at significantly lower margins)."

## NEW KID ON THE BLOCK

In July 1966, MGM Records in New York introduced a micro cassette tape player at a distributors' meeting held inside the Waldorf Astoria. Representing a challenge to 4 Track and 8 Track, the so-called PlayTape cartridge measured 3 3/4 by 2 1/4 inches, with music or voices recorded on one-eighth-inch magnetic tape. New Yorker Frank Stanton, a real estate investor who got his start in the business world trading surplus military goods following World War II, developed the PlayTape system. He said the device was aimed at teenagers eager for a small, portable tape player that would readily record. The product fit in nicely with a boom in transistor radios that was sweeping the nation. Each tape could carry about four songs, and a 30-minute tape was in the works. He also saw potential for the recording device in offices, mostly for dictation. The least expensive player would retail for $29.95, with each cassette selling for $1.29.

In August, Telepro's patent infringement suit against Lear Jet was proceeding forward with witness testimony. The case centered on whether Lear's system was based on TelePro's patent, which was issued in 1957. Lear denied the claim, stating that his 8 Track system was based on a different setup in which the player was fully automatic. The case was being heard in a federal courthouse in Wichita, Kan., where Lear Jet was based. Crucial to the case would be whether the defendant could prove his system worked on a different principle.

Prior to Labor Day, Finley announced ITCC had established a cartridge distribution center and warehouse in Los Angeles, primarily to service the West Coast markets of California, Oregon, and Washington. It was the first move into the backyard of Muntz, who at the time controlled 80 percent of the 4 Track market in southern California. But those numbers dropped considerably in more northern locations — 42 percent in San Francisco, and 20 percent in Seattle. The data showed advertising among the Big Four automakers was driving business for the 8 Track. The ITCC plant would handle cartridges from a Detroit duplicating plant, with the first batch representing 200,000 units for the fall season. ITCC also showed off its new offices in New York, which for the first time consolidated executive, sales, billing, and bookkeeping departments under one roof. The space also included a large showroom where visitors could test various tape players, cartridges, and speakers.

Tired of foot-dragging among some in the record industry, Finley challenged the retail trade to get behind cartridge players. Attending the National Association of Record Merchandisers' (NARM) mid-annual meeting at Chicago's Continental Plaza Hotel, Finley sensed caution among rack-jobbers in distributing cartridges to hundreds of record stores and retail outlets. Amos Heilicher, NARM's secretary, told *Billboard's* Ray Brack that the industry was "on the verge of a tremendous opportunity with tape cartridges." He also pointed out that the market had yet to reach maturity. "We feel we should put ourselves in the best possible position to service the market," he said, "but at the same time, we want it done on a sound basis. We want to avoid excessive and costly obsolescence at the very beginning."

For his part, Finley tried to coax the rack-jobbers into the marketplace, using a mixture of competition and the prospect of new profits. "The automotive and parts people are far more knowledgeable than record people about tape cartridges," he said in the same article. "Record merchandisers have had this dirty word 'tape' bothering them for some time. Now, 70 percent of our tape cartridge business is in the nonrecord merchandising field. If record people don't watch out, the cartridge business will slip away from them. It's starting to already."

In September, Muntz celebrated Stereo-Pak's five-year anniversary. In a special advertising section, company employees waxed poetic about "the most colorful U.S. business personality of the past three decades (who) rolls head-on into his fifth Stereo-Pak year securely installed as the field's most celebrated figure." Still, the entrepreneur was careful to separate his name from his past business selling inexpensive, portable televisions. At several points in the special ad section, it was noted that Muntz Stereo-Pak Inc. was not affiliated with Muntz TV.

The section also touted Muntz' planned 12 Track player and his disdain for 8 Track. "We once thought it might be economical to produce 8 Track machines, because we could save on tape," Muntz said. "You can actually build an 8 Track unit for less than a 4 Track. As it turned out, the tape savings did not nullify the aggravations of the 8 Track unit, especially the position of the pinch wheel in the cartridge. I repeat, if a compatible unit (12 Track) is available, no one is going to buy 8 Track."

At the same time, RCA, Ford, Motorola, Lear, Columbia Records, and others professed their loyalty to 8 Track. "Our basic strategy was to use the automobile to unlock the home market," said RCA's Tarr at the NARM conference. "Now all four major automobile manufacturers, plus Volkswagen, will have 8 Track tape players as optional new car equipment this fall. This is not a trend; it is a tide."

Motorola's Kusisto also reminded conference attendees that while 4 Track enjoyed strong market share along the West Coast, there was little presence of the product in the Midwest, the South, and along the East Coast. "Although we developed a 4 Track player in 1956 and are still tooled to produce it, we wanted major music firms to show their hand," he said. "Meanwhile, Ford wanted sufficient time capability on a cartridge to accommodate an entire Broadway show. When this became possible (with the development of the 8 Track), Ford got excited. We were called in and got the contract. Originally they wanted a compatible unit, but we talked them out of it (as a combined 4 Track and 8 Track player would not have been ready to meet the launch of Ford's 1966 car line)."

Between speeches and panels, Muntz spread the news that he was setting up a tape division in Chicago in 1967. Finley also worked the floor, stating that ITCC had established a third distribution facility, this one in Cincinnati, to service Midwest accounts. "Like Macy's and Gimbels, we do not tell each other what the other is doing, and we do not agree about many things such as 4 Track versus 8 Track, " Finley wrote following the conference. "Muntz says 4 Track; we say 8 Track and 4 Track. One thing we do agree on is the stereo tape cartridge concept is here to stay and will become the most important aspect ever in the music industry."

As the cartridge industry expanded into Europe, *Billboard* was itself the subject of an article in early October. The Soviet Union was using the trade publication's reports on the industry as a guide in establishing the medium. While the Cold War was being waged, the Soviet Union wasn't necessarily blind to a good thing. At first, cartridge players were to be installed inside large tractor-trailers, followed by the consumer market. Unlike the U.S., the Soviet Union had few radio stations. To provide entertainment to long-haul drivers, the country installed short-wave radios inside trucks. Trouble was, once the sets were installed, the truckers typically tuned into foreign broadcasts — Voice of America or the BBC. By providing pre-recorded cartridges, the Soviet Union intended to make available to its drivers "a potpourri of light music, party propaganda, and educational material," said an unnamed staff member of the Soviet Embassy.

Back home, Ford announced it was opening a laboratory to study the "customizing" of vehicle sound systems near its headquarters. "Radio test and development has largely been the province of our radio suppliers, who have done an outstanding job in improving components through the years," said Herbert L. Misch, a Ford vice president. "We feel, however, that further improvements in system reliability call for us going much deeper into this area ourselves." The development work, he told *Billboard*, could lead to tighter specs for Ford's radio suppliers as well as improved sound relative to specific vehicle interiors.

Eager to stay ahead, Muntz announced plans for a global expansion. With the Japanese supply base running at full steam, it wasn't clear where the company would source players. "The move indicates a policy switch from merely looking at the foreign market as supplementary sales through several exporters to looking at the world as one market," said the newly hired president of Stereo-Pak's international division, Don Gordon, then 27.

On the 8 Track front, Elvis Presley's early love for the medium generated an exclusive RCA promotion in the fall of 1966 — a full-color photo of the singer, signed in his handwriting: "I hope you like the new Stereo 8, Sincerely, Elvis

Presley." The promotion included three double albums. In turn, other artists released on 8 Track at that time included Louis Armstrong, The Monkees, and the original cast of the Lincoln Center's production of "Show Boat," among more than a dozen other introductions. RCA also launched foreign language instruction 8 Track cartridges in partnership with the Institute of Language Study. French, Spanish, German, and Italian made up the first of what would become a dozen offerings.

As the 1966 holiday season opened, Muntz offered a gift. His 4 Track cartridges would drop $1 to $4.98, the result of greater efficiencies achieved at a new plant in Van Nuys, he said. The lower price, scheduled to be introduced in January, was offered immediately as a discount. Dealers were free to participate in the program. He explained the sudden drop as a quick way to sell existing tapes marked at $5.98. "We are now able to produce cartridges at a much lower price than before, as a result of our new facility," he told *Billboard*, "and in line with our tradition, we always pass our manufacturing savings along to the consumer."

Record companies and the retail trade were swift to denounce the price drop, insisting Muntz lacked the authority to arbitrarily drop prices. Citing a contractual right, Muntz said he was allowed to sell cartridges as low as $5, and that $4.98 was technically correct. Dot Records, MGM Records, and several other companies said the move "could be detrimental to the success of the cartridge industry." Insisting the campaign was not intended to ignite a price war against 8 Track, Muntz said, "There is no contest between the two systems. We are so far ahead of them now, they'll never catch up." The competition held firm: 8 Track cartridges were almost uniformly sold for $6.98.

As a result, retailers and record stores began promoting 8 Track aggressively. Higher profit margins for the cartridges, and a large advertising program for the holiday season, along with a rash of new releases, further encouraged merchants. Finley, reporting in early December, said the 8 Track was outselling the 4 Track by a four-to-one margin. "Most significant are the sales figures from southern California, where 4 Track is gradually losing ground to the 8 Track field," he wrote. "Up to this point, 4 Track has comprised almost 80 percent of the volume, but today's report for southern California shows that 4 Track accounts for 66 percent and 8 Track 34 percent." Overall, according to ITCC, there were 700 releases on 8 Track and more than 2,000 on 4 Track among 76 record labels in the U.S. By February 1967, ITCC reported 8 Track was outselling 4 Track by eight to one.

Pushing ahead for his plans for a Chicago factory and distribution facility, Muntz spelled out more details of his Midwest and East Coast expansion. The

25,000-square-foot duplication plant, to be staffed by 125 people, would produce up to 15,000 4 Track cartridges a day when fully operational.

Advances were arriving on several other avenues. 3M touted a new lubricated magnetic tape that would improve performance "under temperature and humidity extremes peculiar to the automobile stereo tape cartridge market," *Billboard* reported in February 1967. Nortronics, meanwhile, introduced a new 8 Track playback head that could record or reproduce voices and music, or a combination of the two, while Capitol Records, in partnership with Gauss Electrophysics in Santa Monica, developed a duplication machine, dubbed G-12, that was up to four times faster than conventional equipment.

A master promoter, Finley described the scene at the Automotive Accessories Manufacturers Association Show held at the New York Coliseum over several days in mid-February 1967. "On Wednesday, the Automatic Radio booth was especially swamped inasmuch as famous personalities such as Enoch Light, Horace McMahon, and Henny Youngman were greeting the public," he wrote. "From 3 p.m. to 4 p.m., Lionel Hampton and his Jazz Inner Circle entertained the crowds in the typical Hampton fashion of showmanship. His performance drew practically everyone from each exhibit, and Hampton's rendition of 'Flying Home' brought cheers from the crowd." He touted a new album by Dinah Shore as well as his appearance, as part of a larger panel, on "Long" John Nebel's radio show on NBC. "We were told that a goodly amount of telegrams and phone calls poured into the NBC studio from people who are interested in this new facet of entertainment," Finley said.

As March began, Capitol's Livingston said the record company would start offering its catalog on 4 Track, in addition to 8 Track. Citing bootlegging, the company said it was losing sales while artists were deprived of royalties. Columbia also began to offer releases on 4 Track. Asked if RCA would follow Capitol and Columbia into 4 Track, Norman Racusin, vice president and general manager, said the company did not want to lead the industry down the primrose path. "We, too, have received substantial monetary offers from those who seek to license the manufacture of our catalog in the 4 Track format," he told *Billboard*. "We have rejected these offers on several grounds. ... Inasmuch as we believe in the superiority and long-range growth of Stereo 8, we can see nothing at this point to justify burdening ourselves, our distributors, and our retailers with costly and unnecessary duplication of our catalog in another cartridge format."

Store owners were largely opposed to the plan. Mindful of the speed wars in the late 1940s, in which Columbia offered 33 1/3 records and RCA 45, dealers weren't eager to carry double inventory. "It's like 1948 all over again," Al Ohnhaus of Hunt's Music House in White Plains, N.Y., told *Billboard*. "If they all decided on 8 Track, why don't they stick to it? A strong stand is needed."

D.R. Krantz of Broadway Music Co. in Salt Lake City was even more emphatic. "This represents quite an investment, one which most dealers can't afford. If we have to carry each album in mono, stereo, open reel, reel-to-reel cartridge, 4 Track, and 8 Track, cartridge stock must, by necessity, be thin. Besides, where can we store and display all those tapes?"

## TWICE AS MUCH

Jim Russell, marketing director for Craig Panorama, a 4 Track manufacturer in Los Angeles, wasn't happy. A series of national radio ads by 8 Track tape producers touted that consumers were getting twice as much music for their money. In fact, if an album was 38 minutes, he noted both 4 Track and 8 Track cartridges offered the same amount of music. He told *Billboard* in April 1967 the higher price for an 8 Track cartridge is an "extra expense a consumer incurs in buying" a unit and cartridges. "We feel the consumer will now decide which system will be the majority choice." In addition, he said the company would begin to produce 8 Track tape players, which, according to Finley, would be marketed to "the entire world."

During a trip to Europe at the time, Finley saw potential for 8 Track tape players in automobiles at first. Setting up joint ventures with European firms, the expansion markets included France, Germany, Italy, Sweden, and the United Kingdom. "With the enthusiasm we have encountered from key executives of Europe's leading companies, we foresee the 8 Track market 'happening' on the Continent within the next two years," he wrote in a cable from London.

Eager to expand east of the Mississippi, Muntz was on the move, making plans to open the first of what he said would be many Cartridge City stores across the U.S. Not wasting time, the first unit located outside of Los Angeles was set in Detroit, which *Billboard* dubbed "8 Track City." Located on the city's east side, the 10,000-square-foot store employed 40 workers who handled up to 100 installations a day. Merchandise was air-shipped from the company's Van Nuys operation. Muntz said his 4 Track players and cartridges were overpriced in Detroit. "Our $4.98 cartridges are going for $5.98 to $6.98, and a unit we put in for $49 is going for as high as $80," he said. "These are some of the reasons why nothing has happened here." The entrepreneur said he was interested in opening a store in Cleveland, while a planned outlet in Chicago was put on hold until the company could find the right location.

The Detroit store included a conveyor system, or assembly line. Once a customer paid for a 4 Track player, his or her car was driven to the back of the store, through a series of bay doors, and into an attached warehouse. Units were installed in as little as six minutes. The four-step process included mounting the player below the dashboard, cutting holes for the speakers, and wiring, followed by final installation. The price for the least-expensive unit was $39.95, although several upgrades were available.

At the same time, Lear made front-page news in late April when he sold Lear Jet Aviation to Gates Rubber Co. in Denver. Lear Jet Stereo was part of the deal. Lear was appointed chairman of the aviation and 8 Track divisions, in a bid to ease any concerns among buyers and suppliers. After investing millions of dollars and countless hours into the development of a corporate jet, Lear was eager to cash in on what had proved to be a runaway success. As the deal was consummated, Campbell was appointed interim vice president of operations. A few months later, he took the position of vice president of Lear Jet Stereo after a replacement had been hired to run Lear Jet Aviation.

Because Lear's initial design for 8 Track included a heavy armature to promote stability, the tape tended to speed up or slow down when a vehicle took a sharp turn. Campbell, in his book, *From the Riverbank to Paradise Valley*, says he was intrigued with taking over the Lear plant in Detroit, "where every business problem known to man existed." Upon arriving in the Motor City, Campbell quickly determined the Lear player was too expensive to produce and suffered from numerous quality issues.

"One only had to look at the pile of distributor returns to know that a very serious problem existed," Campbell wrote. "The first step taken was to spend one Saturday putting all (of the) units returned out in the employee parking lot, where we had two large industrial rollers smash all (of) the units. These units were then loaded into a large dump truck which took them to a dump, where previous arrangements had been made to bury them."

Rebuilding the sales department, Campbell got approval from Gates to rehire the original sales manager. Leading a redesign of the player, Campbell and his new team introduced the unit at the National Distributors meeting in New York City at the Waldorf Astoria. At the meeting, Lear was presented with a gift from Gates Rubber — a gold cartridge and 8 Track tape player set on a wood stand.

Next up was the fight for royalties. Because an attorney with Gates Rubber believed the patents "weren't worth the paper they were printed on," Campbell took it upon himself to collect royalties, extending the task of master licensee to JVC of Japan. Had the patent attorney pursued the matter more aggressively, a longer-term deal would have been signed, Campbell said. The patents were set to expire in 1975. Everyone agreed to pay the license fee, save for Pioneer. Arranging a meeting with the company's top brass in Japan, Campbell said the encounter was cordial, but far from positive.

"In those days, the best gift you could give a Japanese executive was either a bottle of Johnnie Walker Black Label or a Cross pen and pencil set," he writes. "With a bottle of Black Label, I made an appointment to meet the owner and president of Pioneer. It didn't take long for him to tell me that they had hired a patent attorney, with his office in Washington, D.C., and he had advised them that our patent wouldn't hold up in (a) United States court. He would not accept my gift."

Upon his return, Campbell received approval to hire an outside patent attorney, who advised the company to wait until Pioneer entered the Denver market. The federal courts on the West Coast were fairly liberal, making it harder to prove damages, Campbell said. It didn't take long for a Pioneer importer, Craig Corp., to enter the Denver market when it moved a batch of players into a warehouse. The company violated four patents, the court determined. The patents dealt with the automatic switch that started the player and, conversely, the radio when the cartridge was removed, the rotating playback head, and the method of holding the cartridge. With the victory in hand, Lear Jet Stereo received an estimated $1 million from Pioneer and Clarion Corp.

"No need to go into more details," Campbell writes. "Lear Jet Stereo won the suit when the judge ruled the patents were valid. ... Shortly thereafter, the president of Pioneer called a meeting between us, to be held in his boardroom. The meeting started with his attorney saying that his advice was for Pioneer to appeal the court's decision, but the owner overruled him. Through the interpreters involved, this perfect gentleman said he wouldn't appeal, because he had his day in court and lost. That is what he wanted, and it was now over. He presented me with a check for over $500,000 in American currency, saying the check represented the most expensive bottle of whiskey he had ever purchased. You had to like the guy."

Under terms of the licensing agreement, JVC would collect 25 percent of all royalties, with the remainder going to Lear Jet Stereo. As things played out, Lear Jet Stereo earned "millions of dollars" from such charges. But by the mid-1970s, with the patents set to expire, Campbell received a call from the patent attorney from Gates Rubber. "He requested that I make a trip to Japan to get the 8 Track license agreements renewed, because their termination dates were fast approaching," Campbell writes. "He mentioned that his head patent attorney had just returned from Japan, but was unable to get the job done. I reminded him that I predicted this. ... Why should any company put out cash when they all had patents to trade?" Try as he did, Campbell traveled to Japan, only to come away empty-handed. The patents soon expired.

At the time Campbell took over at Lear Jet Stereo, Robert Fickes, president of Philco-Ford, hinted at a national convention in Las Vegas in June 1967, that the automaker was contemplating offering a cassette player for the 1969 model year. The cassette player was, after many false starts, starting to gain traction, he noted. Sound quality, playing time, size, the ability to record, and ease of operation were improving. The so-called dual-hub music cassette players — developed by Philips of Holland in 1963 and first introduced in America in 1965 at the Consumer Electronics Show — were projected to sell some 300,000 units in 1967. Trouble was, Philips requested that potential licensees agree to special fees and royalty payments from the outset, which delayed the U.S. debut. When the company decided to waive all of its fee and royalty claims in exchange for an agreement that licensees follow Philips' standards, the market took off. "Had the cassette been able to provide all of the safeguards that 8 Track offered prior to our launch (in October 1965), we would have went with the cassette," King says.

By comparison, sales of 8 Track players totaled 900,000 units from October 1965 through June 1967. Motorola's Kusisto predicted another 400,000 units would be sold by the end of 1967, with a million units projected for the following year.

Behind the scenes, the cassette announcement was premature. Around the time of Fickes' address, Frey asked King, who had since been promoted to product design engineer and would soon move up to system engineer for radio/stereo/electronic communications, to arrange a "Sound Off" between 8 Track and cassette. Using the latest Motorola unit and a new Philips cassette player, King ordered a master tape of Paul Mauriat, a French orchestra leader and conductor of Le Grand Orchestre de Paul Mauriat. "With Frey and some of the other Ford executives, I played the exact same music using 8 Track and the cassette," King says. "We used the latest equipment of the era and the 8 Track was judged to have higher fidelity. The cassette required that a user take out the tape and flip it over to hear the other side. The feeling was that (that) action, along with the other limitations, wasn't enough to bring (the) cassette to market. Flipping the tape was too much of a distraction for drivers."

Mindful that cartridge sales were generating interest among a range of listeners, the record industry began to move away from mono sound. In mid-1967, manufacturers started offering consumers mono and stereo records for the same price, while other labels dropped mono prices by as much as 50 percent in a bid to reduce inventory. Dave Rothfeld, division merchandising manager at E.J. Korvette, a chain of discount stores centered in the East and Midwest, told *Billboard* that an album by The Monkees was selling 2-1 in favor of mono, but when the prices were equalized, sales were 7-1 in favor of stereo. Other dealers reported similar results. Popular records showed the most gains, while classical music was largely unaffected, given fans had long since switched to stereo.

An added benefit for cartridge sales came from the used-car business. As cars equipped with players made it to used-car lots, a whole new set of cartridge-buyers emerged. Meanwhile, variety cartridges, featuring multiple vocalists rather than a single album by an artist, were selling briskly. Tarr said RCA planned to take out additional ads to drive 8 Track sales in the lead-up to the 1967 holiday season, in publications including *Life, Time, Esquire, Sports Illustrated, Newsweek*, various auto magazines, and trade publications like *Cash Box*. RCA also enlisted celebrities like retired pitcher Sandy Koufax, one of the most loved players in baseball before he retired at the age of 31 in 1966, to move the 8 Track. RCA co-sponsored the popular "Sandy Koufax Show" on the NBC Radio network.

Retailers also played a part in boosting sales. Frank Meckrock, owner of a line of Radio Frank stores, said a "Please Handle" sign at cash registers served to boost sales. "First, we tried keeping the cartridges (behind) the counter, but now we put them out front where the customers can handle them," Meckrock told *Billboard*. "First thing you know, the customer is picking up two or three cartridges he never intended to buy." He added that if a clerk suspected a customer might swipe a cartridge from the counter, they were instructed to provide "extra special service and attention." At Mobile Stereo in Cleveland, which operated three stores, a campaign to distribute 100,000 placemats advertising tape players and cartridges inside restaurants proved to be a hit. "It's a direct method," Charles Lombardo, the owner, told *Billboard*. "They're presented in an educational and informative manner. They serve to acquaint the consumer with 4 and 8 Track cartridges."

At the Consumer Electronics Show, held in late June in New York, the innovation that turned the most heads was a fast-forward button that would propel an 8 Track cartridge at four times normal speed, introduced by Lear Jet Stereo. The company also showed off Synchro-track and pitch control to eliminate crosstalk. Universal Tapedex, meanwhile, offered a competing fast-forward selection that moved tape at 10 times normal speed.

Others speculated the cartridge sector had yet to tap its potential. "I think the industry should be doing some basic research to determine how broad the market can be," said Jim Tiedjens, president of Midwestern Tape Distributor in Milwaukee, in *Billboard*. "We're a mobile society now so we have to take the merchandise where the people are, and they're in hotels, motels, resorts, filling stations, tollway oases, and dozens of other locations. ... For example, if you're in the West, you will find tapes explaining the Grand Canyon and you will play these as you come into the area."

**HEAD TO HEAD**
Early 8 Track tape player
playback heads were
used to produce sound.

# ALL GROWN UP

## RISE OF THE CLUBS

INTENT ON OFFERING A CATALOG OF popular music across all media, the record companies began to align their releases at the start of 1968. Most of Columbia's cartridge releases, for example, trailed that of an album by roughly 30 days, while other record companies had as large of a gap as 90 days. The so-called simultaneous release era was under way, vowed Elliot Horne, RCA's recorded tape product planning manager. "We're rapidly narrowing the gap," he told *Billboard*. "We believe that the hit albums are basically the hit cartridges, so we have to move fast." By March, the trade publication reported 179 of the top 200 albums were available on cartridge, with 8 Track leading the way.

To drive demand, ITCC initiated a sales contest among merchandisers. Selling 8 Track cartridges in 100-pack and 30-pack shipments, the distributor offered "top name, top label" albums. For the first five months of 1968, the cartridges were available at a suggested retail list price of $2.99 (regular cartridge pricing ranged from $4.95 to $6.95). The dealers who sold the most tapes and set up display windows using materials provided by ITCC would win various prizes. In the first month, the promotion racked up an additional $4 million in sales. "If you are a dealer, ITCC would like you to understand that your distributor was not negligent in getting shipments to you, as sales exceeded the ITCC projections to the point where only a portion of the merchandise ordered was able to be shipped," Finley wrote. He vowed that every dealer would be able "to participate in the most fabulous contest ever held in the history of the music industry."

The distributor also concluded a deal in March with TWA, whereby 8 Track cartridges would be played at the departure gates in eight airports, including San Francisco, Los Angeles, Chicago, Boston, New York, and Washington, D.C. One hour prior to takeoff, airport personnel would roll up a cart equipped with an 8 Track tape player and speakers. Once the music started, TWA officials reported passengers at the departure gates were dancing, humming along, and tapping their feet. TWA also installed 8 Track tape players in its Ambassador Clubs. Other airlines like Continental and Olympic quickly followed suit.

Another distribution offering was tape cartridge clubs, which typically offered lower prices and promotional items such as cigarette lighters, cleaner kits, watches, and cameras. The Ford Stereo 8 Club offered incentives to every buyer who purchased a new car equipped with a player, including $1 off each cartridge purchased. The Tape Club of America, meanwhile, offered its members one-third off the list price of a cartridge. A free gift, typically a cleaner kit, was sent to anyone who signed up to be a member through a mail-in advertisement form found in popular magazines like *Hot Rod*, *High Fidelity*, and *Car Life*.

With some 10 million cartridges sold since late 1965, tape duplicators and distributors decided to pass on special promotions for the 1968 holiday season. "The industry is at a point where consumer demand exceeds the supply," said Jim Johnson, sales manager of Ampex. "We don't believe there is any need for a special seasonal promotion at this time." He added the company was operating at full production with three daily shifts. Other record companies, meanwhile, packaged and sold Christmas releases featuring one or multiple artists, including Motown Records. The holiday selections included such artists as Diana Ross and The Supremes, Marvin Gaye, Smokey Robinson and The Miracles, and the Four Tops.

On the cassette front, sales advanced among younger buyers, but there was pushback. When Ampex placed an ad in *Playboy* in November 1968 that touted its cassette recorder could record directly off the air, not to mention records and 4 Track and 8 Track cartridges, the company immediately downplayed the feature. Facing significant criticism, E.P. Larmer, Ampex's vice president and general manager, issued a statement.

"We are in a very competitive market," he said. "Cassette equipment like the

Ampex Micro 30 is now being sold by a dozen companies and more will soon enter the field. This will be an important product in the consumer market, and as one of the first major companies to sell it, we intend to claim our share of sales. In the future, however, we will change the emphasis of our advertisements to cover other features of the Micro 30, but will continue to make some mention of recording off the air as a technical capability of the product. We don't feel that the introduction of this type of equipment is going to hurt sales of pre-recorded music, in which our company also has an important stake."

In an editorial, "Playing With Fire," published in early December, *Billboard* denounced the action of radio stations that openly encouraged the recording of music. In one instance, a New York DJ told listeners to get their tape recorders ready for 10 p.m. that evening when the station would play a new Beatles LP, the so-called White Album. "This is mischief of the most devastating sort, and it is a sample of what is happening in many stations across the nation," the editorial read. "It symbolizes a policy which is at once immoral and unethical; a policy which can corrode the very fabric of the record-music industry. Ultimately, radio itself will suffer, inasmuch as its health is largely dependent on the prosperity of the record-music industry."

Al Berman, president of the Harry Fox Agency in New York, which collected fees for music publishers, was more direct. Citing the practice of taping radio programs or record albums, Berman said it was "illegal and morally reprehensible." The company's practice, according to *Billboard*, was to send a letter to the advertising agency that created a print or on-air advertisement that promoted such recordings. The letter outlined the nature of the illegality and stated that the agency could be held liable if the practice continued. They also threatened legal action if the ads continued.

Behind the backdrop of the cassette controversy, the cartridge industry was racking up unprecedented growth. The pre-recorded tape industry had generated close to $250 million in sales in 1968, and most projected 1969 would dwarf the prior year's track record. Propelling projections for tape cartridges, which in three years accounted for 25 percent of the total dollar volume of recorded music, was a move by E.J. Korvette to install new display racks that allowed consumers to handle cartridges while greatly limiting theft. Under the new program, tape cartridges were wrapped in clear plastic around a large cardboard panel. Two packages, set side by side, would fit into a regular display rack for records. What's more, numerous record outlets and department stores were replacing monaural records with cartridges.

On the marketing front, Boston-based Automatic Radio launched a national advertising campaign in early 1969 that included print ads in major magazines, along with a commercial message presented by Johnny Carson and Ed McMahon from the popular "Johnny Carson Show." The ads would reach 4 million viewers in some 2.5 million homes. Another promotional campaign was tied to "The Newlywed Game," which aired on ABC on Saturday nights. The Walt Disney Co. also entered the cartridge market, licensing such titles as "Mary Poppins," "Snow White," "Jungle Book," "Peter Pan," "Bambi," "Winnie the Pooh," and "Cinderella." The offerings, as well as those from Warner Bros. and others, proved to be popular among parents keen on keeping their children entertained on long road trips.

Eager to maintain retail prices while driving down production costs, numerous cartridge manufacturers were moving their assembly operations to Mexico, near the U.S. border. With labor roughly one-fifth the cost in Mexico as compared to the U.S., based on union wages, nearly every major record company and tape manufacturer had operations in Tijuana, Mexico City, or Mexicali. In almost all instances, cartridge parts were shipped from the U.S. to plants in Mexico. Following final assembly and return shipment, the cartridges were subject to a U.S. duty based on "value-added," meaning the cost of Mexican labor. It was that "compelling economic consideration" that drove Lear Jet Stereo to move its headquarters and tape operation from Detroit to Nogales, Ariz., as well as Sonora, Mexico, Campbell says.

Beyond home and auto units, 8 Track manufacturers reported what was initially a niche market for portable players had blossomed by the summer of 1969. "There are no blue skies in the statement that our business in portables is more than three times better this year than last year," said Jim Gall, marketing vice president at Lear Jet Stereo, in *Billboard*. "We are anticipating an eventual four-time growth factor before the year is up." Ed Mason, president of Belair Enterprises, was even more enthused. "The portable player is on fire," he said. "We had sales of about $4 million in the year ending on March 31, and we expect our sales to jump between $15 million and $20 million this fiscal year."

## QUAD-8

In 1970, Muntz' Stereo-Pak was feeling the weight of several setbacks, including a major fire the previous year that had caused an estimated $600,000 in damages at the company's Van Nuys operation. A resulting fight with the insurance company, which finally agreed to pay half of the damages, had slowed production and distribution of 4 Track players and cartridges considerably. During the previous summer, General Recorded Tape announced it had stopped manufacturing 4 Track, which resulted in further depletion of the market. Muntz, in a bid to boost sales, opened what he referred to as "instant stores" — essentially sales trailers offering demos and 4 Track cartridges that would be parked near high schools, colleges, and concert venues. But unless consumers had a 4 Track player, few were interested in buying the cartridges.

Unable to keep pace with advertising, marketing, and promotions, 4 Track was further weakened by what Muntz foresaw as a growing problem: namely, bootleggers. With fewer hit albums available on 4 Track, some in the record industry were cherry-picking what they could. Capitol Records, having signed a deal with Ampex, left Muntz without one of his main sources of material. Others in the 4 Track sector simply dropped their cartridge prices, depleted their inventory, and got out of the market. Others placed more emphasis on 8 Track and cassette sales. While Chrysler hinted it might offer an optional cassette player in its 1971 vehicles, GM and Ford held off from committing to a timetable. Unless the cassette was made to flip automatically, the auto giants weren't interested.

"The cassette system is like going back to 4 Track," Shari Smith, manager of a Stereo City store outside of Chicago, told *Billboard*. "The sound isn't as good as 8 Track. As for the record feature, how many people want to record in the car? The market is limited when you talk about the record feature — mainly doctors and salesmen who want a coordinated system and come in asking about cassettes." She added 8 Track cartridges were outselling cassettes by a 10-to-1 margin.

By May, Stereo-Pak was reformed as Muntz Stereo Corp. of America, with Muntz selling his interest. The new entity, sans Muntz, focused its product line on hardware, including AM/FM radios and tape players rather than pre-recorded music. The new company, which would drop the Muntz name in favor of Clarion in 1975, made its public debut at the Consumer Electronics Show in New York in late June. By the end of 1970 it rolled out a cassette adapter for 8 Track players, as well as a reversible cassette player. With Muntz out of the picture, Muntz Stereo Corp. embraced the 8 Track like never before. At the same time, production had solidified on two fronts — Japan was the world's largest supplier of 8 Track players, while Mexico was shaping up as the epicenter of cartridge assembly.

As the 4 Track faded from the spotlight, 8 Track and cassette systems were each enjoying success. To improve the experience of listening to the 8 Track, Motorola, Lear Jet Stereo, Ford, and RCA, among others, rolled out Quad-8. The quadrasonic system used four channels of sound that produced signals that were independent of one another. To demonstrate the system, the industry developed a marketing tape where a single voice would travel independently from one speaker to the next — right channel front, left channel front, right channel rear, and left channel rear. Also referred to as "surround sound," Motorola was intent on making Quad-8 available on 8 Track systems. "There is a high probability that (Quad-8) will be offered as a factory or dealer-installed option in 1972 or 1973 models," Kusisto told *Billboard* in mid-1970. Left unsettled at the time was which system — 8 Track, cassette, or reel-to-reel — would be the first to roll out Quad-8.

In 1967, the British rock group Pink Floyd presented the world's first quadra-sonic concert at Queen Elizabeth Hall in London, using custom speakers and equipment the rock band had developed. By late 1970, RCA released some 40 Quad-8 tapes, each priced at $7.95, $1 more than regular cartridges (players were roughly 25 percent higher per unit). "The use of a new, slightly thinner tape combined with recent economies in tape-coating will permit the marketing of Quad-8 cartridges at only a modest premium over conventional cartridges," RCA's Tarr told *Billboard*. He said Stereo-8 and Quad-8 could be played on con-ventional players, though the surround sound experience would only work with a compatible playback unit. In addition to RCA, Capitol Records, Liberty, and United Artists introduced Quad-8 tapes, while Columbia took a wait-and-see approach. Capitol, which first distributed John Lennon's "Imagine" album on Quad-8, introduced the upgrade to radio stations on an experimental basis.

The new offering fit in with what consumers were looking for in the industry — more music selections, better sound, longer play, and ease of use. In addition to Quad-8, the industry embraced an emerging noise reduction system — which eliminated the "hiss" common in pre-recorded music at the time — developed by Ray Dolby and his Dolby Laboratories. For more variety, Qatron Corp. introduced a fully automatic, fully programmable 8 Track stereo changer for the home, with a base price of $299. The round unit, about the size of a hatbox, held 12 cartridges and played each automatically. If a consumer desired more variety, the first channel of each program could be played in sequence. There was even a repeat function that could play a single cartridge again and again.

At the Consumer Electronics Show in 1970, in addition to the rollout of the new Muntz concern, numerous manufacturers showed off 8 Track tape record-ers, including Panasonic, Craig, Sony, and Lear Jet Stereo. The emergence took away a key component of cassette players, although the 8 Track hadn't really suffered from the lack of a record function based on revenue activity. Despite a national recession, sales of 8 Track players and cartridges surged. Cassette sales also blossomed. In essence, the 8 Track got a late jump on recording, while the cassette lacked pre-recorded music at the start. As the two systems began to compete on equal footing, slight differences were magnified in various advertising and marketing efforts. The cartridge was twice as fast in playback speed, and offered greater fidelity and better quality (cassette tapes had a greater tendency to jam), while the cassette was smaller, easier to store, and offered more varying lengths for recording. The late start for cassette in pre-recorded music affected the medium more than 8 Track's lack of recording. According to *Billboard*, based on industry sources interviewed during the Consumer Electronics Show, blank cassettes were outselling pre-recorded cassettes by a 10-to-1 margin. On the flip side, an automatic rewinding function had been introduced, which eliminated the need to remove a cassette and turn it over, thereby removing a safety problem in the eyes of the automakers. The only problem with these types of players was that they were expensive.

The 1969-70 recession ushered in another challenge, especially on the 8 Track front. The insurance industry reported an alarming rate of tape player thefts from automobiles, with some insurers considering a specific ban on covering stolen units. To prevent coverage loss while enhancing its bottom line, National Tape Distributors came out with a novel program: The distributor of pre-recorded cartridges allowed insurance companies to process claims through National's catalog. In turn, customer claims would be processed by National's nearest facility. Among the replacement players available to a policyholder, one model came with a built-in burglar alarm.

In an effort to better serve the industry, Kusisto, Finley, and Campbell, among others, formed the International Tape Association (ITA). The nonprofit organization, which would have its first convention in early 1971 at the Astroworld Hotel in Houston, would track statistical data, credit information, and sales volumes; provide consumer and trade education, and public relations; and promote copyright protection. "ITA will not seek to compete with the other trade groups, but will strive to work with them toward building a stronger and more stable tape industry here and throughout the world," said Finley, who was named ITA's executive director. Kusisto was named chairman of ITA's executive committee. The association predicted 8 Track would be a $1.5-billion industry by the end of the 1970 calendar year.

As the industry ramped up promotions for the holiday season, car sales slowed across the board. Compounding the situation, General Motors was dealing with a national strike by the UAW, which began in October. The automaker hinted it would offer cassette players as an option should production resume in short order. At the same time, Chrysler, via Motorola, planned to offer cassette players as an option in a handful of 1971 model year vehicles. Department stores like Montgomery Ward & Co. were riding the tape bandwagon. The retailer rolled out a "young image" campaign in its 360 stores that encouraged teenagers and young adults to visit its newly revamped music departments, with the hope that other merchandise would get a boost.

As 1971 unfolded, Electrodyne Corp. in Hollywood unveiled a jukebox-like device called Record-A-Tape, which went on to become Make-A-Tape. Holding 150 master tapes, the tape-duplicating center, roughly the size of a jukebox, allowed consumers to punch in the album of their choice and, within four minutes, walk away with an 8 Track tape cartridge loaded with music. A single cartridge cost 85 cents, with a pre-recorded album available at $2.49 to $3 ($3.49 for a double album, using the same cartridge). For storeowners, the $4,500 machine offered several advantages over pre-recorded cartridges, namely no returns, no obsolete stock, no pilferage, and no warehouse overhead.

## IS IT LIVE, OR IS IT MEMOREX?

If one commercial could transform an industry dogged by quality problems, Memorex was living proof. Founded in 1961 in Silicon Valley, Memorex Corp. was initially a supplier of precision magnetic tape to the computer industry. A decade later, the company, founded by Larry Spitters and three engineers, entered the precision cassette tape market. A graduate of Western Michigan University, the University of Michigan Law School, and Harvard University, Spitters worked with the Leo Burnett ad agency to develop, at first, a commercial touting the company's premium line of cassette tapes. The early TV ads showed an opera singer shattering a glass with a high note. The effect was then repeated using a Memorex tape. While technically a human voice could break a thin wine glass, the rare effect was best achieved through amplification.

In 1972, Memorex brought in jazz singer Ella Fitzgerald and re-shot the commercial. After shattering the glass during a studio performance, a commercial announcer relayed the news that it would be repeated, only this time with a recording. As the glass broke into pieces, the announcer intoned, "Is it live, or is it Memorex?"

Not an industry to leave a boast unchallenged, Audio Magnetics Corp. in Gardena, Calif., touted its sales leadership in a series of print ads. "While our competitors are busy breaking glass, we're breaking records." The ad went on to say that nine out of 10 of America's largest retailers favored Audio's cassettes (as listed in *Fortune* magazine). The print ad went on to close, "Remember, it all depends on whether you want to sweep up broken glass or clean up at the cash register."

While Memorex admittedly relied on amplification to break each glass, to the average consumer, the ad was compelling and memorable. But more importantly, it helped drive revenue for the entire industry. While pre-recorded cassette tapes lagged, sales of blank cassettes were at an all-time high. Annual cassette sales in the U.S. — from Ampex, Maxell, 3M, BASF, Norelco, Memorex, and Audio Magnetics — topped $116 million in 1971, $90 million of which was blank tapes.

On a dollar basis, combined cassette tape sales reached $100 million in 1971, up 20 percent from the previous year. The growth was even more remarkable with the nation still in the midst of a recession. At the same time, industry leaders moved to rid the world of "cheapies" — poorly built and sourced cassettes. "Our collective reputation among consumers is being damaged by shoddy cassettes being produced by opportunistic fast-buck operators," Paul B. Nelson Jr., vice president and general manger of North American Philips' home entertainment products division, told *Billboard*.

Overall, sales of records accounted for 55 percent of the pre-recorded music market in 1971, with 8 Track representing nearly 35 percent and the balance going to cassettes, according to various record companies.

As the year closed, word leaked that Ford would offer Quad-8 players as an option for the 1972 model year. The players, to be supplied by Motorola, were to be marketed to consumers and automotive dealers simultaneously. To that end, Motorola would supply dealers who answered a survey with a Quad-8 display equipped with a player and four speakers. Just as in the early days of 8 Track, Motorola was out to prime the market. "Too many dealers are just as confused as consumers when it comes to explaining (Quad-8)," said Ken Thomson, manager of Motorola's consumer products division, in *Billboard*. He added that quadrasonic would take longer to develop "if software producers continue their foot-dragging."

Retailers like Sears, J.C. Penney, and Montgomery Ward were picking up the message. Radio Shack, with 1,200 stores across the country, set up Quad-8 listening areas in each of its outlets. Founded in Boston in 1921 by two brothers, Milton and Theodore Deutschmann, Radio Shack at first provided equipment to ham radio operators — the store name was a reference to small, wooden shacks that housed radio equipment on ships. At different points throughout its history, the retailer, which today has more than 4,000 stores, manufactured its own products, including a line of blank cassette tapes it assembled at a plant in Fort Worth, Texas, starting in 1972.

In an unusual retail setting, two New York businessmen opened an audio products store inside a decommissioned Douglas DC-7 on Long Island near Roosevelt Field. The store, called Fly By Nite, included two stereo sound rooms, a Quad-8 listening booth, space for compact radio equipment, service call buttons, an accessory area, and a mini radio station located inside the cockpit. The wings of the plane would offer a stage for concerts, with the tail providing a backdrop for advertising slide images. The plan was to set up other planes in 15 cities, including Baltimore, Boston, Washington, Philadelphia, and Atlanta.

Kusisto, who noted that 8 Track sales had reached "a magical plateau," said 8 Track, combined with Quad-8, "will prove to be a very difficult combination to beat." By the end of the year, with some 24 million 8 Track tape players now inside cars and homes, he said Japanese suppliers would deliver 12 million units annually. Still, the move to Quad-8 was becoming a story of too much too soon. While the upgrade was an important product in the grand scheme of consumer electronics, Campbell noted the number of models shown at the Consumer Electronics Show in June was overwhelming. "I feel it is a necessary item, but I have a feeling that the main reason so many audio tape people are carrying them is that they are afraid not to carry one," he said. "Nobody really wants to be caught off guard."

Further, Campbell noted some new features, like a fast-forward function, weren't necessary. "It's nice to have, but you don't have to have it," he said. "When you come right down to it, the track selector is almost as good. If a feature pays for itself and is good for the customer, then it's a good feature. If it doesn't, and in some cases fast-forward in 8 Track doesn't, then it is of no help to the manufacturer or retailer to have the feature. For instance, fine-tuning is not that necessary. Obviously a unit is fine-tuned when it is purchased. If the consumer fools around with it then he's going to knock it out and have to bring it back, and this just means money spent that didn't have to be spent. Think about it. Who is going to sell a unit out of tune?"

For the first time, Lear Jet Stereo marketed a line of cassette players for the car. Campbell said the company wasn't abandoning the 8 Track; rather, it was moving to specialize in sound systems for use in automobiles and the home.

In May 1974, *Billboard* reported that Ford would offer a Motorola-built Quad-8 as an option in three 1976 models — Lincoln Continental, Lincoln Mark IV, and Ford Thunderbird. The player, Kusisto said, would come equipped with receivers capable of handling discrete and matrix playback offerings. One of the first Quad-8 cartridges to be offered was "The Best of Diana Ross" on Motown Records. RCA, which partnered with Motorola to develop Quad-8 in 1970, said it would offer 50 titles initially to support surround sound. Other companies came forward to provide software, including Columbia Magnetics, a CBS subsidiary that unveiled a Quad-8 cartridge that could record and play back sound, as well as Panasonic, Fisher Radio, and General Electric. Overseas, Quad-8 players were selling briskly, especially in Europe and Asia, where consumers appreciated high-end music players. By late summer 1975, a quarter of the most popular 200 albums listed on *Billboard's* Top LPs and Tapes chart were available in Quad-8.

To celebrate the 10th anniversary of the 8 Track, the International Tape Association, which billed itself as the world's largest international audio and video trade group, invited King to speak at its October 1975 seminar at the Marriott Essex House in New York City. In his address, "Four-Channel Tape, The Next Step Forward in Automotive Tape Systems," the Ford engineer noted that since the introduction of the 8 Track, "many companies entered the tape cartridge hardware and/or software industry, and we believe that all the efforts, coupled with overwhelming customer acceptance of (8 Track) in the automobile and home, have resulted in developing the stereo tape industry into the 'giant' that it is today." He added that the introduction of the Quad-8 player was the next step forward in the industry.

Following the speech, Finley, as ITA's executive director, along with John Jackson, chairman of the seminar and a member of the association's advisory board, presented King with a large plaque. The award read: "Presented to the Ford Motor Co. in recognition of the 10th anniversary of their introduction of 8 Track stereo in automobiles. Their imagination and confidence led the way for others to follow. Ford's success in creating consumer awareness of this innovative entertainment medium led to the development of the home market. Because of the Ford Motor Co., our industry has prospered and profited by the sale of tape decks, magnetic tape, and cartridge parts." It ended with the phrase, "Ford, a major contributor to making tape a household word throughout the world."

By the time King made his speech, he had been promoted to executive engineer of fuel, exhaust, tires, and frames within Ford's powertrain and chassis division. When he was named an executive engineer in 1973, he no longer had direct oversight of the 8 Track program. He went on to spend 32 years at Ford, retiring in 1997 as regional manager of Asia Pacific and new markets for Ford's Customer Service Division.

Recalling the early days of 8 Track, King says the competition was intense, but friendly for the most part. "I do recall early after the introduction of 8 Track, when Don Frey — who was vice president in charge of the Ford Division at the time — asked if Motorola could mount a couple of speakers in the engine compartment of his car, just behind the front grille. Don had an 8 Track cartridge that would play Ford's latest commercial, which was 'Ford has a better idea.' Well, Don's neighbor was Pete Estes, who was a vice president at GM. Every morning, just as Don was pulling out of his driveway, he would play the tape really loud as he passed by Pete Estes' house."

As King oversaw other programs at Ford, 8 Track slowly lost ground to cassette players. The reasons included the expiration of the Lear patents, which caused the industry to compete on price. As a result, manufacturers installed cheaper parts, and quality suffered. In turn, children who had grown accustomed to cassette players in the late 1960s and early 1970s preferred the medium to the 8 Track by the time they reached driving age. Cassette players also were lighter, less expensive, and used less power. The oil embargo in 1973-74, along with the adoption of CAFÉ standards by Congress in 1975, resulted in smaller cars and trucks. The two acts caused more consumers to switch from large, luxury vehicles to lighter models — more and more of which came outfitted with cassette players.

In addition, record stores were keen to clear out cartridges, given cassettes took up less space. With the added room, merchants could expand their prerecorded offerings. "I don't think anyone saw what the 8 Track industry would become," King says. "Things were moving so fast, but as I look back it was quite an accomplishment. Ford, Motorola, RCA, and Lear set aside their competing interests for the greater good of delivering a product that would benefit consumers. Had we each clung to our individual interests, the industry would have been much different. I am neither sad nor upset that 8 Track was supplemented by cassette, which in turn was supplemented by compact discs, and then downloads. What I'm most proud of is that the four companies came together and worked as a team. That is the true measure of success, in my mind."

The last major release on 8 Track was Fleetwood Mac's "Greatest Hits" (Nov. 1988).

The cartridge era ended, as it began, on a high note.

# ABOUT THE TEAM

## JOSEF BASTIAN, ADVISOR

Josef Bastian is a thought leader who, since 1991, has specialized in broad-based workforce development, education, and training initiatives. An internationally published author and poet, Bastian's works include *Nain Rouge, Nain Rouge II: The Red Tide, The Red Truth, A Pardon for Vincent, Somewhere in Middle America, Next Halloween, Beyond the Little Brauhaus, Big Boss Man, Keith Doodle Saves Christmas, Middleness,* and *A Day in the Life of Denver Penny.*

He also served as writing consultant for *Passport to the Corner Office: The Starter's Guide to Corporate Life,* written by R.J. King. Bastian holds advanced degrees in instructional design and learning technology from Oakland University in Rochester Hills, Mich.; the University of Detroit Mercy in Detroit; and Oxford University in England. He can be reached at *jbastian67@msn.com.*

## CASSIDY ZOBL, DESIGNER

Cassidy Zobl is an award-winning graphic designer and art director. From 2007 to 2014, she served as art director of *Hour Detroit,* a successful lifestyle magazine that serves more than 46,000 monthly readers in metro Detroit. In addition, she has created multimedia marketing campaigns, program books, playbills, posters, signage, logos, and promotional material for a diverse range of businesses and individuals. She also served as art director of *Passport to the Corner Office: The Starter's Guide to Corporate Life,* written by R.J. King.

As a designer and art director, Zobl has been honored with multiple gold and silver medals, including best feature design in *FOLIO* magazine's national competition (2011), feature spread design by the Society of Professional Journalists (2010, 2012), and the General Excellence category from the City and Regional Magazine Association (2009), the organization's top award. She also won an Ozzie for "Best Use of Photography" for a fashion feature (2012). For more information, visit *cassidyzobl.com* or *cassidyz@gmail.com.*

## ANNE BERRY DAUGHERTY, COPY EDITOR

Anne Berry Daugherty received a Bachelor of Arts degree in advertising from Michigan State University in East Lansing, Mich. She has worked as a writer and editor for Sandy Corp., The Quarton Group, and Crain Communications. She is currently a freelance copy editor with a variety of clients in metro Detroit, including the Community Foundation for Southeast Michigan. She does work for publications such as *DBusiness* magazine, *Hour Detroit* magazine, *Detroit Home* magazine, *Carolinas Golf, Metropolitan Detroit Dining Guide,* and *Metropolitan Detroit Guest and Resource Guide.* She also served as copy editor for *Passport to the Corner Office: The Starter's Guide to Corporate Life,* written by R.J. King. She can be reached at *jd2794@aol.com.*

## ABOUT THE AUTHOR

For more than 20 years, R.J. King has covered one of the nation's busiest news towns. He has generated award-winning coverage as editor and co-founder of *DBusiness* magazine and *DBusiness Daily News*, and as a business writer at *The Detroit News*. He also had his first book, *Passport to the Corner Office: The Starter's Guide to Corporate Life*, published in September 2013. Mr. King has written more than 4,000 articles and interviewed hundreds of business owners, CEOs, entrepreneurs, scholars, artists, and politicians. He also attends 200-plus events annually, including the North American International Auto Show, industry conferences, cultural ceremonies, and charitable fundraisers.

King is a board member of Beyond Basics, the Asian Pacific American Chamber of Commerce, the Detroit Athletic Club Executives Club, Detroit Aircraft Corp., Brazen Sports, and the Brother Rice Business Alliance. He also serves on the board of trustees of The Parade Co.

*DBusiness*, launched in 2006, has garnered 24 editorial gold and silver medals from the Alliance of Area Business Publishers (2010-2016), including being named the top business magazine among 70 regional business publications in the United States, Canada, Puerto Rico, and Australia, which collectively reach more than 1.2 million professionals. Over the course of 16 years at *The Detroit News*, Mr. King traveled in Europe and Asia, where he covered international business and political leaders. Back home in Detroit, he broke hundreds of stories such as Ford Motor Co.'s $2-billion transformation of the historic Rouge Industrial Complex in Dearborn, Mich., into a model of sustainable manufacturing.

King grew up in Bloomfield Village, a suburb of Detroit, where he was the middle child: three older sisters and a brother, and three younger sisters and a brother. An alumnus of the University of Michigan-Dearborn, King lives in a historic neighborhood in Detroit.

For inquiries: *rj@rjkingpublishing.com*.